T0092730

Internet Tiered Services
Theory, Economics and Quality of Service

George N. Rouskas

Internet Tiered Services
Theory, Economics, and Quality of Service

 Springer

George N. Rouskas
North Carolina State University
Raleigh, NC
USA

ISBN: 978-0-387-09737-4   e-ISBN: 978-0-387-09738-1
DOI: 10.1007/978-0-387-09738-1

Library of Congress Control Number: 2008943579

Printed on acid-free paper.

springer.com

To Magdalini, for always being there

Preface

As telecommunications products and services have become an essential part of everyday life, consumers have at the same time grown intimately familiar with the concept of tiered pricing that is associated with such services. With tiered service structures, users may select from a small set of tiers that offer progressively higher levels of service with a corresponding increase in price. Tiered structures have been applied in several forms to wireless services (e.g., characterized by the amount of voice minutes, number of text messages, or the size of one's circle of friends to whom voice calls are free), Internet broadband access (e.g., the access speed or volume of monthly transferred data), and digital TV offerings (e.g., the number of channels included), among others. Service tiering is a form of market segmentation which, if applied appropriately, benefits both providers and consumers by making available services and associated price points that reflect the diversity in consumers' needs and ability to pay.

The purpose of this book is to develop a theoretical framework for reasoning about and pricing Internet tiered services, as well as a practical algorithmic toolset for network providers to develop customized menus of service offerings. We provide a comprehensive study of the design, sizing, and pricing of tiered structures for Internet services, and we illustrate their potential in simplifying the operation of complex components such as packet schedulers. The topic corresponds to a graduate-level field of study that combines the fields of Internet services, economics, and quality of service (QoS) in network resource allocation.

This book is intended for practicing engineers as well as industry and academic researchers. The potential audience includes: network designers and planners, and engineering and sales managers at Internet Service Providers involved in the selection, sizing, and pricing of tiered services, including bundled services; industry practitioners and graduate students in computer science, telecommunications, and related fields interested in Internet services and economics; and researchers who wish to explore the subject matter further. The book is also suitable as textbook for graduate-level courses in electrical engineering, computer engineering, computer science, operations research, and economics programs, that address Internet services, Internet economics and pricing, discrete location theory, or dynamic programming optimization, as well as for industry short courses in these areas.

Book Organization

The book provides a comprehensive treatment of the problems and issues arising in providing Internet tiered services. Chapter 1 (Introduction) provides a motivation for, and definition of, tiered network services, and discusses existing business models in this context. The remainder of the book is divided into three parts, each addressing a distinct aspect of tiered services.

Part I, *Theory,* consists of Chapters 2-6. This part builds upon concepts from location theory to develop a theoretical framework for reasoning about and tackling algorithmically several important problems related to tiered network services. Chapter 2 provides an easy entry to facility location problems, and introduces a new variant, the directional p-median problem, as the fundamental problem underlying the study of tiered services. It also discusses several applications of this problem in large-scale, heterogeneous networking and computing environments.

Chapters 3 through 5 investigate the problem of optimal tier selection for services characterized by a single parameter, namely, the data rate of the access link; this problem is modeled as a directional p-median problem on the real line (i.e., one dimension). Chapter 3 considers the case where user demands are deterministic and known in advance to the network provider, and presents efficient optimal algorithms for determining the tiers to be offered so as to minimize the cost, in terms of network resources, incurred by the provider. Chapter 4 introduces the concept of "TDM emulation" that refers to a tiered-service network in which all service tiers are multiples of a basic bandwidth unit. TDM emulation is useful in several network contexts, including next generation SONET/SDH networks and traffic grooming. The tier selection problem is formulated as a constrained version of the directional p-median problem, and a suite of efficient algorithms is presented to determine jointly the basic bandwidth unit and service tiers in a near-optimal manner. Chapter 5 provides approximate yet accurate solutions to the one-dimensional directional p-median problem with stochastic demands, i.e, when only a probabilistic distribution of user demands is known.

Chapter 6 concludes the first part of the book by considering the directional p-median problem in multiple dimensions; this problem arises naturally whenever the network (1) provides a service characterized by multiple parameters, or (2) bundles several distinct services together as a single product. The tier selection problem in this case is shown to be NP-hard, and several heuristic algorithms are presented and evaluated. The main conclusions from the work presented in Part I is that a small set of appropriately selected service tiers is sufficient to approximate the resource usage of continuous-rate networks.

Part II, *Economics,* contains Chapters 7-9, and employs concepts from economic theory to formulate the problem of tier selection in a manner that takes into account the realities of the marketplace. Chapter 7 considers a bandwidth tiered service and develops an economic model that considers three perspectives: that of users only, that of providers only, and that of both simultaneously. It also uses game-theoretic techniques to determine optimal prices for each service tier in a manner that balances the conflicting objectives of users and providers. Chapter 8 investigates tiered service as a market segmentation strategy for increasing provider profits under the assumption that consumer behavior with respect to pricing differs across the user population. Chapter 9 extends this study to service bundles, and uses tools from economics and utility theory to determine optimal service tiers when customers face budget constraints. An interesting finding from Part II is that simple tiering structures currently deployed by service providers may not work well in practice.

Part III, *Quality of Service*, illustrates the practical implications of tiered services by considering a central component of packet-switched networks, the link scheduler. Chapter 10 presents a survey of the literature on packet scheduling disciplines, and discusses the trade-offs between simplicity and ease of hardware implementation, one the one hand, and the requirement for delay and fairness guarantees, on the other hand, facing the designers of schedulers for high-speed routers. Chapter 11 then demonstrates that using the principles of service tiering it is possible to design a new scheduler, referred to as tiered service fair queueing (TSFQ), that achieves tight delay bounds and worst-case fairness with low algorithmic and implementation complexity. Experimental results from a Linux kernel implementation of the TSFQ scheduler are also presented.

Raleigh, North Carolina, USA   *George N. Rouskas*
October 2008   rouskas@ncsu.edu
                                                                   http://rouskas.csc.ncsu.edu/

Acknowledgments

The material in this book is for the most part the product of a multi-year effort within my research group at North Carolina State University. Therefore, this book would not have been possible without the contributions of the following former graduate students. Laura E. Jackson introduced and defined the directional p-median problem (Chapter 2) and developed algorithms for deterministic (Chapter 3) and stochastic demands (Chapter 5) and for the multi-dimensional problem (Chapter 6). Nikhil Baradwaj designed a more efficient algorithm for deterministic demands (Chapter 3) and formulated and solved the constrained directional p-median problem (Chapter 4). Qian Lv developed the economic models for tiered services (Chapters 7, 8, and 9). Zyad Dwekat made contributions to the design of the tiered service fair queueing (TSFQ) scheduler, Ajay Babu Amudala Bhasker developed a simulation model of TSFQ in *ns-2*, and Shrikrishna Khare created an implementation of the scheduler in the Linux kernel (Chapter 11).

Portions of the book were reprinted with permission from IEEE from the following articles: Qian Lv, George N. Rouskas, "An Economic Model for Pricing Tiered Network Services." In *Proceedings of IEEE ICC 2009*, June 14-18, 2009, Dresden, Germany; George N. Rouskas, Nikhil Baradwaj, "On Bandwidth Tiered Service." *IEEE/ACM Transactions on Networking*; Qian Lv, George N. Rouskas, "On Optimal Sizing of Tiered Network Services." In *Proceedings of IEEE Infocom 2008 Miniconference*, April 13-18, 2008, Phoenix, Arizona; George N. Rouskas, Nikhil Baradwaj, "TDM Emulation in Packet-Switched Networks." In *Proceedings of IEEE ICC 2007*, June 24-27, 2007, Glasgow, Scotland; George N. Rouskas, Zyad Dwekat, "A Practical and Efficient Implementation of WF^{2}Q+." In *Proceedings of IEEE ICC 2007*, June 24-27, 2007, Glasgow, Scotland; George N. Rouskas, Nikhil Baradwaj, "A Framework for Tiered Service in MPLS Networks." In *Proceedings of IEEE Infocom 2007*, pp. 1577-1585, May 6-12, 2007, Anchorage, Alaska; Laura Jackson, George N. Rouskas, "Optimal Quantization of Periodic Task Requests on Multiple Identical Processors." *IEEE Transactions on Parallel and Distributed Systems,* vol. 14, no. 8, pp. 795-806, August 2003; and with permission from Elsevier from the following article: Laura E. Jackson, George N. Rouskas, Matthias F.M. Stallmann, "The Directional p-Median Problem: Definition, Complexity, and Algorithms." *European Journal of Operational Research*, vol. 179, no. 3, pp. 1097-1108, June 2007.

I also want to thank the National Science Foundation for supporting my research over the past dozen years, and in particular for grants CNS-0322107 and CNS-0434975 which led directly to the research results reported in this book.

Many people have influenced my professional career, but none more so than Mostafa Ammar, my former advisor at Georgia Tech, who has served as a role model since our first meeting in September, 1989. I am grateful for his guidance and for instilling in me a love and appreciation of fundamental research.

I also thank Springer for the opportunity to publish this book, and in particular Alex Greene and Katie Stanne for their help and support throughout the book-writing process.

Finally, I wish to thank my wife Magdalini for her constant encouragement and patience all these years, especially during the final months of writing.

Contents

Acronyms

ADSL	Asymmetric digital subscriber line
ATM	Asynchronous transfer mode
BB	Bimodal-bimodal demand distribution combination
BSFQ	Bin sort fair queueing
BSDH	Bidirectional supply driven heuristic
CAIDA	Cooperative Association for Internet Data Analysis
CDF	Cumulative distribution function
CS	Concentration set
CSP	Constrained shortest path
DAG	Directed acyclic graph
DDH	Demand driven heuristic
DH	Decomposition heuristic
DPM1	Directional p-median problem on the real line (one-dimensional)
DPM2	Directional p-median problem on the plane (two-dimensional)
DRR	Deficit round-robin
FCFS	First-come, first-served
FI	Fairness index
FIFO	First-in, first-out
FR	Frame relay
FRR	Fair round-robin
GB	Gigabyte
GE	Gigabit Ethernet
GFP	Generic framing procedure
GPS	Generalized processor sharing
GRIA	Global/regional interchange algorithm
GSM	Global system for mobile communications
HC	Heuristic concentration
IP	Internet protocol
ISP	Internet service provider
JDPM1	Directional p-median problem on the real line, variant for joint optimization of number and location of supply points

Kbps	Kilobit per second
LCAS	Link capacity-adjustment scheme
LFVC	Leap forward virtual clock
LSP	Label switched path
MAX-ES	Maximization of expected surplus problem
MAX-S	Maximization of (provider) surplus problem
Mbps	Megabits per second
MPLS	Multi-protocol label switching
NBR	Normalized bandwidth requirement
NES	Normalized expected surplus
OC	Optical carrier
OPT-P	price optimization problem
PDF	Probability distribution (density) function
PM	p-median problem
PM1	p-median problem on the real line (one-dimensional)
PM2	p-median problem on the plane (two-dimensional)
PTH	Power of two heuristic
PON	Passive optical network
QoS	Quality of service
QQ	Quadrimodal-quadrimodal demand distribution combination
RNG	Random number generator
RR	Round-robin
SCFQ	Self-clocked fair queueing
SDH	Synchronous digital hierarchy
SDNAP	San Diego network access point
SDPM1	Stochastic directional p-median problem on the real line
SEFF	smallest eligible virtual finish time first
SFF	smallest virtual finish time first
SLA	Service level agreement
SONET	Synchronous optical network
SRR	Smoothed round-robin
S-RR	Stratified round-robin
STFQ	Start-time fair queueing
STS	Synchronous transport signal
TB	Teitz & Bart heuristic
TCP	Transmission control protocol
TDM	Time division multiplexing
TDM-DPM1	Directional p-median problem, variant for TDM emulation
TSFQ	Tiered service fair queueing
UB	Uniform-bimodal demand distribution combination
UDP	User datagram protocol
USDH	Unidirectional supply driven heuristic
UU	Uniform-uniform demand distribution combination
VC	Virtual clock
VCAT	Virtual concatenation

VoIP	Voice over IP
VPN	Virtual private network
WFI	Worst-case fairness index
WFQ	Weighted fair queueing
WF^{2}Q	Worst-case fair weighted fair queueing
WRR	Weighted round-robin

Chapter 1
Introduction

Over the last decade, the growth and uses of computer networking have increased at an explosive rate, and the global Internet has become an infrastructure of importance equal to that of the power grid and the transportation system. The success of the Internet has to do with the advances in the underlying technology as well as the innovations in the services available to users. The last few years have witnessed dramatic and continuing improvements in networking technology, including optical and wireless transmission systems and link capacities, router and switch speeds, and network protocol and software capabilities. These improvements have made it possible to operate and manage effectively networks that may support performance differentiation and guarantees in serving heterogeneous users with a wide range of requirements. As the Internet has evolved into a ubiquitous and reliable communication system, it has changed the way people interact, access information, or conduct business.

Computer networking has enabled new and exciting applications that have transformed every aspect of business and commerce, education and scientific exploration, entertainment and social interactions, and government and military services. The increased reliance on, and utility of, computer networks has created an insatiable user demand for better and faster Internet connectivity and services, and an entire industry has emerged to develop related products. Whereas workplace and University networked environments have traditionally provided Internet access to employees, researchers, educators, and students, recent years have seen a proliferation of home networks with high-speed broadband access, public wireless hot-spots, and smartphones with Web browsing and data networking capabilities that further extend the reach of the Internet in terms of both geography and user diversity.

Technology advances typically lead to the emergence of new business models and service offerings, which in turn influence the direction for further research and development in the underlying technologies, and so on. This interplay between technology and business models creates a virtuous cycle of innovation of which users are the primary beneficiaries. Interestingly, however, there has been a notable discrepancy between the designers and developers of networking hardware and software included in the equipment making up the Internet infrastructure, on the one hand,

G.N. Rouskas, *Internet Tiered Services*, DOI: 10.1007/978-0-387-09738-1_1,
© Springer Science + Business Media, LLC 2009

and the Internet service providers (ISPs) that devise the products to make available to customers, on the other. This discrepancy manifests itself in terms of the *granularity* at which users may access the services offered by the network. Specifically, networks are typically designed with fine granularity in mind, whereas ISPs market products developed around models of coarser-grained services, as we explain in the following.

1.1 Continuous-Rate Packet-Switched Networks

Historically, packet-switched computer networks, including the Internet and legacy networks based on Asynchronous Transfer Mode (ATM) or Frame Relay (FR) technologies, are designed to be *continuous-rate*. In a continuous-rate network, users may request any rate of service (bandwidth), and the network must be able to accommodate arbitrary requests. Theoretically speaking, continuous-rate networks may allocate bandwidth at very fine granularities; for instance, one client may request a rate of 98.99 Megabit per second (Mbps), while another customer may ask for 99.01 Mbps. Taken to the limit, bandwidth in such networks could potentially be allocated at increments of 1 bit per second (bps). Clearly, the option of requesting arbitrary rates offers clients maximum flexibility in utilizing the available network capacity.

On the other hand, supporting bandwidth allocation at such extremely fine granularity may seriously complicate the operation and management of the network. Based on the above example, the network provider faces the problem of designing mechanisms to distinguish between the two rates (i.e., 98.99 Mbps vs. 99.01 Mbps) and enforce them in an accurate and reliable manner. However, the task of differentiating between the two users on the basis of these two rates may be extremely difficult, or even impossible to accomplish for traffic of finite duration, undermining the network's ability to support important functions such as robust traffic policing or accurate customer billing. Furthermore, given the unpredictability of future bandwidth demands in terms of their size, arrival time, and duration, link capacity across a continuous-rate network may become fragmented. Such fragmentation poses significant challenges in terms of traffic engineering, and may compromise the ability to achieve an acceptable level of utilization or meet users' quality of service (QoS) requirements.

To illustrate the challenges associated with operating a continuous-rate network, consider packet fair scheduling algorithms such as the weighted fair queueing (WFQ) [90] or its variants. The WFQ discipline can be used to allocate the capacity of a link to any number of competing flows in proportion to the weights assigned to each flow, as well as to guarantee a strict upper bound on the delay of each packet of a flow under certain conditions. However, existing implementations of the WFQ discipline or its variants have been designed under the assumption that flow weights can be arbitrary; in other words, they are designed to allocate the link bandwidth at the finest possible granularity. As a result, these implementations suffer from severe scalability challenges which have impeded their wide adoption in Internet routers. We discuss packet fair queueing disciplines in more detail in Part III of the book, where we present a scalable implementation based on the concept of tiered services that is introduced next.

1.2 Tiered-Service Networks

Due to the issues involved in making fine-granular services available to a large, heterogeneous user population cost-effectively, in practice, most network operators have developed a variety of *tiered service* models in which users may select only from a small set of service *tiers* (levels) which offer progressively higher rates (bandwidth). The main motivation for offering such a service is to simplify a wide range of core functions (including network management and equipment configuration, traffic engineering, service level agreements, billing, and customer support), enabling the providers to scale their operations to hundreds of thousands or millions of customers. Returning to the previous example, a tiered-service network might assign both users requesting 98.99 Mbps and 99.01 Mbps to the next higher available rate, say, 100 Mbps. In this case, there is no need to handle the two customers' traffic differently; furthermore, the network operator only needs to supply policing mechanisms for a small set of rates, independent of the number of users.

For a more formal definition, consider a network that offers a service characterized by a single parameter, e.g., the bandwidth of the user's access link. A tiered-service network is one that offers p levels (tiers) of service, where typically p is a small integer, much smaller than the number n of (potential) network users (i.e., $p \ll n$). Let

$$Z =< z_1, z_2, \cdots, z_p > \tag{1.1}$$

denote the vector of service tiers offered by the network provider. Without loss of generality, we make the assumption that the service tiers are distinct and are labeled such that

$$z_1 < z_2 < \cdots < z_p. \tag{1.2}$$

Users wishing to receive service are limited to only these p tiers, and may subscribe to any tier depending on their needs and their willingness to pay the corresponding service fee. In particular, z_1 is the minimum and z_p the maximum amount of service that a user may receive. In the case of residential Internet access, for instance, z_1 may correspond to a minimum bandwidth for the service to be considered "broadband," while z_p may correspond to the capacity of the access link, e.g., as determined by limitations imposed by Asymmetric Digital Subscriber Line (ADSL) technology.

According to this definition, traditional telephone networks and transport networks based on Synchronous Optical Network or Synchronous Digital Hierarchy (SONET/SDH) technology [43] belong to the class of tiered-service networks. Indeed, such networks allocate bandwidth in discrete tiers that are multiples of a basic unit rate that corresponds to the slot size in the underlying Time Division Multiplexing (TDM) system.

Motivated by the discrete nature of bandwidth allocation in TDM systems, an early study by Lea and Alyatama [71] investigated the benefits of "bandwidth quantization" in packet-switched networks. In their terminology, "bandwidth quantization" refers to sampling the (effectively) continuous range of possible rates to select a small set of discrete bandwidth levels that are made available to users; in essence, these levels correspond to the service tiers we defined earlier. This work was carried out with the goal of reducing the number of states required to analyze broadband ATM networks under the assumption of Poisson traffic, and it presented a heuristic based on simulated annealing to obtain a sub-optimal set of discrete bandwidth levels of service[1]. Because of the significantly reduced state space, it is possible to apply elegant and exact theoretical models to analyze the performance of a tiered-service network efficiently. The main contribution of this study was to demonstrate for the first time that this benefit comes almost for free, as even with a sub-optimal set of tiers the performance degradation (e.g., in terms of call blocking) compared to a continuous-rate network is negligible.

Current tiered service offerings by major ISPs can be broadly classified in two categories based on the tiering structure. The structure of one class of service tiers for Internet access, especially those targeted to business customers, is based on the bandwidth hierarchy of the underlying transport network infrastructure (e.g., DS-1, DS-3, OC-3, etc.), While this is a natural arrangement for the service provider, it is unlikely that hierarchical rates designed decades ago for voice traffic would be a good match for today's business data applications. The second class employs *exponential tiering* structures in which each tier offers twice the bandwidth of the previous one. The various ADSL tiers (e.g., 384 Kbps, 768 Kbps, 1.5 Mbps, 3 Mbps, 6 Mbps, etc.) available through several ISPs are an example of such an exponential structure. While such simple tier structures may be an appropriate choice for marketing purposes, the relationship between these exponentially increasing levels of service (and their price) and the usage patterns (and willingness or ability to pay) of the population of potential subscribers is open to debate.

So far, we have discussed tiering in the context of broadband Internet access services that are generally characterized by the bandwidth available on the customer's access link. However, the concept of tiering is equally applicable to other parameters that may characterize the QoS experienced by a customer's traffic and may be included in the service level agreement (SLA) negotiated between the customer and provider. Consider, for instance, a provider offering a service that guarantees an upper bound on the delay experienced by its customers' packets. On the one hand, the provider is unlikely to be able to support fine-granularity delay bounds (e.g., at the level of nanoseconds) within a network of realistic size even with the most sophisticated (and expensive) QoS mechanisms. On the other hand, users are unlikely to require (or afford) delay bounds at such a level of precision. A more reasonable approach would be to offer a small set of delay bound tiers that are tailored to specific

[1] In Chapter 3 we show that the problem considered by Lea and Alyatama in [71] can in fact be solved optimally, and we present an efficient algorithm for obtaining an optimal solution.

applications, e.g., voice, (stored) video on demand, (live) video conferencing, etc. Such delay bounds are likely to be tied to human perception abilities that allow for coarse granularities of tens of milliseconds, making it unnecessary for the network to have to distinguish packet delays at extremely fine precision.

Similar observations apply to other QoS parameters, e.g., level of protection of user traffic. In this case, it may be possible to define and offer a set of discrete grades of service from which customers may select based on the level of quality of protection [41] appropriate for their traffic. Tiered structures may be also be employed when the offered service is not characterized by bandwidth but by amount of traffic generated. As an example, in early 2008, Time-Warner, a major cable ISP in the United States, started a pilot program in Beaumont, Texas, under which it charges customers based on how much data they transfer (i.e., upload *and* download) [50,51]. For the pilot program, Time Warner put in place an exponential structure with tiers at 5 GB, 10 GB, 20 GB, and 40 GB of monthly traffic.

1.3 Multi-Tiered Pricing Schemes

ISPs around the world have introduced several forms of tiered services along the lines of the model we described above, with each tier associated with a higher level (amount) of service than the previous one with a corresponding increase in price. Multi-tiered price systems are prevalent for both business and residential Internet access, and arise naturally under both pricing schemes, capacity-based or usage-sensitive, that are prevalent for Internet services [62].

1. **Capacity-Based Pricing.** Capacity-based schemes relate pricing to usage by setting a price based on the bandwidth or speed of the user's connection link. This is accomplished by charging for the configuration (i.e., bandwidth) of the connection, but not the actual bits sent or received. This scheme relates to the tiered service model as follows: the service is characterized by the amount of access bandwidth, each of the service tiers $z_1, \cdots, z_p$, corresponds to a certain access speed, and users are charged based on the tier to which they have subscribed. Capacity-based pricing is the prevailing pricing policy for residential broadband Internet access services, although the pilot program by Time-Warner [50, 51] may be the beginning of a shift towards usage-sensitive pricing for residential markets.

2. **Usage-Sensitive Pricing.** Usage-sensitive pricing policies charge users for the actual amount of traffic they generate, hence price is a function of the actual bytes transferred over a certain time period, usually one month. In current practice, ISPs charge business customers (e.g., a video-on-demand provider) based on their traffic volume using a *95-th percentile rule* [52, 114][2] designed to disregard the low probability peak-load periods. Specifically, the ISP measures the

---

[2] Time-Warner's pilot program discussed above is an exception as the ISP charges based on the total amount of bytes uploaded or downloaded, not the 95-th percentile.

user's traffic volume over 5-minute intervals during each billing period (e.g., one month), and charges the user based on the 95-th percentile value among these measured values. Typically, ISPs have a tiered pricing structure [114] in which each of the service tiers $z_1, \cdots, z_p$, corresponds to a certain traffic volume and higher tiers are mapped to higher prices. Such a structure can be mapped to the tiered service model by considering a customer with a 95-th percentile value x such that $z_{j-1} < x \leq z_j$ as having "subscribed" to tier z_j and charging the customer accordingly.

Note that with capacity-based pricing, the tier (e.g., access speed) to which a user subscribes does not change over time (except, for instance, when a user upgrades to a higher speed). With usage-sensitive pricing, on the other hand, a user may be charged according to a different tier every billing period, i.e., depending on the actual traffic volume generated during each period.

If designed and applied appropriately, tiered services and corresponding multi-tiered pricing schemes have the potential to be a catalyst for Internet service innovation and penetration. On the provider side, tiered structures can be an effective tool for ISPs to optimize and specialize their offerings so as to capitalize on the increasing sophistication and requirements of various segments of Internet users, as well as to differentiate themselves from the competition. On the user side, tiered pricing is likely to spur adoption by providing a wide menu of customized services from which users may select based on needs and affordability. To realize this potential, it is crucial that both the service tiers and the corresponding prices be determined in a manner that takes into account simultaneously the (usually conflicting) objectives of users and providers. The purpose of this book is to provide insights into the selection and pricing of tiered structures for Internet services and offer solutions that consider the perspectives of both users and ISPs.

Part I
Theory

Chapter 2
The Directional p-Median Problem: Definition and Applications

In this chapter we first review the classical p-median problem that arises in a wide range of facility location and clustering applications. We then introduce a new variant of the problem, the "directional" p-median problem defined under a new notion of distance, the directional rectilinear distance metric. We discuss the complexity of the p-median and directional p-median problems, and we describe several applications of the latter problem. The directional p-median problem provides the theoretical foundations for studying the network tiered service that is the subject of this book.

2.1 The p-Median Problem

The p-median problem is generally expressed as:

> Given n "demand" points, find p "supply" points that minimize the sum of the distance from each demand point to its closest supply point, with respect to a particular distance metric.

The choice of distance measure impacts the complexity of the problem as well as the approach needed to find a solution, as we discuss shortly. The p-median problem under the Euclidean distance measure has been in existence since at least the 17th century, when Pierre de Fermat posed the 1-median problem with 3 demand points [70]:

> Given three points in the plane, find a fourth point such that the sum of its distances to the three given points is a minimum.

However, the origin of the problem is a matter of debate. For a historical review of the 1-median problem (also referred to as "Weber" problem, after Alfred Weber), the reader is referred to [31].

At the start of the twentieth century, in one of the founding texts in location theory, Alfred Weber considered a version of the Euclidean 1-median problem to deter-

G.N. Rouskas, *Internet Tiered Services*, DOI: 10.1007/978-0-387-09738-1_2,
© Springer Science + Business Media, LLC 2009

mine industrial location while minimizing transport cost [115]. Writing in 1977, Ostresh noted that the problem "has application to the siting of steel mills and schools,
[...] and hospitals" [89]. More recently, *p*-median has arisen in areas such as statistical cluster analysis [68], spatial data mining [38], and data compression.

2.1.1 Continuous vs. Discrete Space

In k-dimensional space, $k \geq 1$, the *continuous* *p*-median problem allows supply
points to be located anywhere in k-space, not merely selected from among the given
demand points. The *discrete* problem, on the other hand, provides a list of *candidate
points* from which supply points may be chosen.

Consider first the *p*-median problem in the plane (i.e., two dimensions) that arises
naturally within the facility location context. Let $D((x_i,y_i),(x_j,y_j))$ be the distance
from point (x_i,y_i) to point (x_j,y_j) according to some distance metric. The Euclidean,
D_e, or the rectilinear, D_r, distance measures are of importance to typical facility
location problems [77], and are defined as:

$$D_e((x_i,y_i),(x_j,y_j)) = \sqrt{(x_i - x_j)^2 + (y_i - y_j)^2} \qquad (2.1)$$

$$D_r((x_i,y_i),(z_j,t_j)) = |x_i - x_j| + |y_i - y_j| \qquad (2.2)$$

The decision version of the continuous *p*-median problem in the plane may be
formally stated as:

Problem 2.1 (Continuous-PM2). Given a set

$$X = \{(x_1,y_1),(x_2,y_2),\ldots,(x_n,y_n)\} \qquad (2.3)$$

of n demand points in the plane, an integer p, and a bound B, does there exist
a set

$$Z = \{(u_1,v_1),(u_2,v_2),\ldots,(u_p,v_p)\} \qquad (2.4)$$

of p supply points such that

$$\sum_{i=1}^{n} \min_{1 \leq j \leq p} \{D((x_i,y_i),(u_j,v_j))\} \leq B ? \qquad (2.5)$$

Under the rectilinear distance measure, it is known that only demand points and
intersection points need be considered as candidates for supply points. Intersection
points are found by crossing the set $\{x_1,x_2,\ldots,x_n\}$ with the set $\{y_1,y_2,\ldots,y_n\}$, and

subtracting the demand points, yielding at most $n^2 - n$ new points. Thus, the continuous *p*-median problem under the rectilinear distance measure reduces to a discrete *p*-median problem.

The discrete *p*-median problem in the plane can be formulated as the following integer program [96]:

Problem 2.2 (Discrete-PM2). Given a set X of demand points in the plane, a set C of candidate points from which a number p of supply points is to be chosen, and d_{ij} the distance from point i to point j, minimize

$$\sum_{i \in X} \sum_{j \in C} d_{ij} r_{ij}$$

subject to

$$\sum_{j \in C} r_{ij} = 1 \qquad \forall i \in X, \tag{2.6}$$

$$r_{ij} \leq s_j \qquad \forall i \in X, j \in C, \tag{2.7}$$

$$\sum_{j \in C} s_j = p, \tag{2.8}$$

$$r_{ij}, s_j \in \{0,1\} \qquad \forall i \in X, j \in C, \tag{2.9}$$

where

$$r_{ij} = \begin{cases} 1 \text{ if point } i \text{ is assigned to candidate } j, \\ 0 \text{ otherwise,} \end{cases} \tag{2.10}$$

$$s_j = \begin{cases} 1 \text{ if candidate } j \text{ is chosen,} \\ 0 \text{ otherwise.} \end{cases} \tag{2.11}$$

The distances between demand points and candidate points are organized into a distance matrix $[d_{ij}]$. In Chapter 6 we discuss properties of the distance matrix that affect the difficulty of finding a good quality solution [104]. The reader is referred to [26] for a complete treatment of discrete location problems.

Fig. 2.1 shows the 2-dimensional *p*-median problem under three different combinations of the solution space (continuous or discrete) and distance measure (Euclidean or rectilinear).

In one dimension, the rectilinear and Euclidean distance measures are the same; hence, the continuous *p*-median problem reduces to a discrete one where the candidate supply points are simply the demand points. The decision version of the *p*-median problem on the line may be formally stated as:

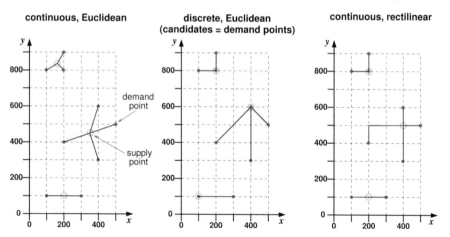

Fig. 2.1 Sample p-median problem instances in the plane, with $n = 9$ demand points and $p = 3$ supply points

Problem 2.3 (PM1). Given a set $X = \{x_1, x_2, \ldots, x_n\}$ of demand points, an integer p, and a bound B, does there exist a set $Z = \{z_1, z_2, \ldots, z_p\}$ of p supply points such that

$$\sum_{i=1}^{n} \min_{1 \le j \le p} \{D(x_i, z_j)\} \le B ? \tag{2.12}$$

2.2 A New Notion of Distance: The Directional Distance Metric

We now introduce the *directional* rectilinear distance measure. In general, a l-directional, k-dimensional rectilinear metric $D_{dr}^{(l,k)}$ (with $1 \le l \le k$) defines distance from point $r = (r_1, \ldots, r_k)$ to point $q = (q_1, \ldots, q_k)$ to be:

$$D_{dr}^{(l,k)}(r, q) = \begin{cases} \infty, & \text{if } r_i > q_i \text{ for some } i \in \{1, \ldots, l\} \\ \sum_{1 \le i \le k} |q_i - r_i|, & \text{otherwise} \end{cases}. \tag{2.13}$$

Thus, in a directional p-median problem, a supply point must *achieve or exceed* the values of the first l coordinates of all the demand points assigned to it. On the real line, this restriction requires that the nearest supply point for a given demand point be located to the right of it, hence, the 1-directional, 1-dimensional rectilinear distance is:

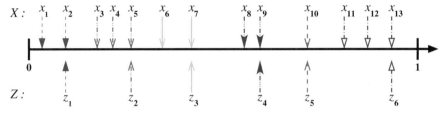

Fig. 2.2 Implied directional mapping of demand points x_i to supply points z_j (© 2007 IEEE)

$$D_{dr}^{(1,1)}(x_i, z_j) = \begin{cases} z_j - x_i, & \text{if } z_j \geq x_i \\ \infty, & \text{otherwise} \end{cases}. \tag{2.14}$$

Let $X = \{x_1, \ldots, x_n\}$ be a set of n demand points on the real line, such that $x_1 \leq x_2 \leq \cdots \leq x_n$. A set of supply points $Z = \{z_1, \ldots, z_p\}$, $z_1 < z_2 < \cdots < z_p$, $1 \leq p \leq n$, is a *directional feasible solution* for X if and only if $x_n \leq z_p$. Associated with a feasible solution is an implied *directional mapping* $h : X \rightarrow Z$, where $h(x_i) = z_j$ if and only if

$$z_{j-1} < x_i \leq z_j. \tag{2.15}$$

Expression (2.15) represents a set of *directionality constraints* that a feasible solution must satisfy. Fig. 2.2 shows the implied directional mapping from a set of 13 demand points onto a set of 6 supply points.

Let n_j be the number of demand points mapped to supply point z_j. The directional p-median problem on the real line (which we will refer to as problem DPM1) can be expressed as a minimization problem as follows:

Problem 2.4 (DPM1). Given a set X of n demand points, $x_1 \leq x_2 \leq \ldots \leq x_n$, find a feasible set Z of p supply points, $z_1 < z_2 < \ldots < z_p$, $1 \leq p \leq n$, which minimizes the following objective function:

$$S(z_1, \ldots, z_p) = \sum_{i=1}^{n} [h(x_i) - x_i] = \sum_{j=1}^{p} (n_j z_j) - \sum_{i=1}^{n} x_i$$

$$= \sum_{j=1}^{p} (n_j z_j) - \rho_X \tag{2.16}$$

where ρ_X is the total demand in input set X.

The objective function (2.16) represents the sum of the distances (under the distance measure $D_{dr}^{(1,1)}$ in (2.14)) from each demand point x_i to its corresponding supply point $h(x_i)$.

In the plane, the 2-directional rectilinear distance is:

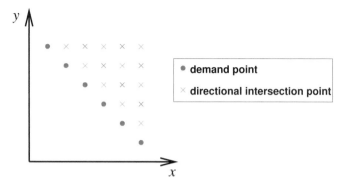

Fig. 2.3 Demand points arranged in a downward slope yield the maximum number of directional intersection points.

$$D_{dr}^{(2,2)}((x_i,y_i),(x_j,y_j)) = \begin{cases} x_j - x_i + y_j - y_i & \text{if } x_j \geq x_i \text{ and } y_j \geq y_i, \\ \infty & \text{otherwise.} \end{cases} \quad (2.17)$$

We define *directional intersection points* to be the subset of intersection points that lie above (at least) one demand point as well as to the right of (at least) one demand point. Specifically, the point (x_j, y_j) is a directional intersection point if:

1. $(x_j, y_j) \notin X$, that is, (x_j, y_j) is not itself a demand point, and
2. there exist points $(x_i, y_i) \in X$ and $(x_k, y_k) \in X$ for which $x_j = x_i$ and $y_j = y_k$ and $x_j > x_k$ and $y_j > y_i$.

Fig. 2.3 shows the worst case scenario, an instance of the directional *p*-median problem on the plane, referred to as DPM2, which has the most directional intersection points possible, $(n^2 - n)/2$.

2.3 Summary of Complexity Results

Hassin and Tamir demonstrate that the one-dimensional *p*-median problem is solved in time $O(pn)$ [46]. In two or more dimensions, *p*-median is an NP-hard problem under either the rectilinear or the Euclidean distance measure [77]. Chapter 6 discusses a few of the many heuristics developed for this problem.

In the next chapter, we present an optimal dynamic programming algorithm for the directional *p*-median problem on the real line. In [59] it was shown that the rectilinear *l*-directional, *k*-dimensional *p*-median problem is NP-hard when $l = k = 2$ (problem DPM2), which implies NP-hardness for all l, k satisfying $2 \leq l \leq k$. We present efficient heuristic algorithms for DPM2 in Chapter 6.

2.4 Applications

The p-median problem arises in domains where the objective is to (1) group elements described by a vector of characteristics, and (2) assign a representative to each group. It was originally conceived in the context of facility location (where the representative element is a facility that serves all the elements of the corresponding group), but it has also been applied to statistical cluster analysis (where elements in the same group are somehow more similar than elements in different groups) and data or vector compression (where the representative is used to describe the whole group).

The directional p-median problem similarly arises in clustering applications in which the representative of each group must be such that for certain characteristics, its values exceed those of the other members of the group (i.e., by imposing directionality constraints similar to (2.15)). Such applications emerge in modern networking and computing environments where catering to very large sets of heterogeneous users/demands poses significant scalability challenges. To overcome these challenges, it is often desirable to arrange users with similar service requirements in a small number of groups, and treat all users within a given group identically by providing each with the same level of service. To ensure that all users receive at least as good a service as they have requested, the service level of a group is determined by the requirements of the most demanding user in that group; hence the directionality constraints in the form of (2.15) arise naturally in such a setting. A practical and effective approach to clustering users in such groups is for the provider to offer a tiered service, i.e., make available only a small set of service levels (tiers) to which customers may subscribe, thus avoiding the complexities associated with supporting the potentially infinite number of service levels that users may request.

Tiered service, and hence the directional p-median problem in its several variants, is useful in several contexts, including, but not limited to:

- **Broadband Internet Access.**
  As we discussed in Chapter 1, most ISPs already offer some form of tiered Internet access service, along with a corresponding multi-tiered price structure that is either capacity-based or usage-sensitive. In most cases, the tiering is either *ad-hoc* or based on simple exponential structures. In the chapters that follow, we develop a formal framework for optimizing the service tier structure in a manner that takes into account the objectives of both users and providers.
- **Dimensioning of MPLS Tunnels for Virtual Private Networks (VPNs).**
  A virtual private network (VPN) is a network that uses a public telecommunication infrastructure, such as a carrier's network, to connect various remote sites of an organization to each other in a secure and efficient manner. Traffic among the various sites is *tunneled* through the public network using various tunneling technologies. Multi-protocol label switching (MPLS) [7, 27, 37, 98] is a widely-used tunneling technology that forwards traffic over what are referred to as label switched paths (LSP) [1, 6].

Typically, the size of (i.e., bandwidth associated with) an LSP (equivalently, VPN tunnel) connecting two remote sites is directly related to the traffic demands between the two sites. In principle, the bandwidth of an LSP may take any value up to the capacity of the physical links over which the LSP is routed. From a traffic engineering point of view, however, supporting LSPs of arbitrary size can be a challenge, and may lead to fragmentation of link capacity, hence low utilization of the underlying network resources. Alternatively, the VPN provider may offer a tiered service in which clients may select from a small, appropriately selected set of LSP sizes (bandwidth levels) [57, 102], the one that best suits their needs. In addition to the benefits regarding the operation and management of the network, such a tiered service also simplifies what in MPLS parlance is referred to as *LSP resizing*, since a client can easily upgrade to the next higher tier whenever its traffic exceeds the current allocation.

- **Efficient Packet Fair Scheduling.**
  In packet-switched networks, the scheduling algorithm is central to the QoS architecture. Timestamp-based schedulers, such as weighted fair queueing (WFQ) [90] and its variants, may be employed to allocate the bandwidth fairly among competing flows, as well as to ensure bounded delay under certain conditions. The main drawback of such schedulers is their high complexity, which makes it difficult to implement in hardware so as to operate at wire speeds. The high complexity is due, to a certain extent, to the fact that these schedulers are designed to accommodate packet flows with arbitrary weights (equivalently, bandwidth shares). Consequently, recent schedulers such as stratified round robin [93] reduce the complexity by arranging flows of arbitrary weights into classes using exponential grouping.
  In Part III of this book we present a new class of schedulers that capitalize on tiered service (as well as certain properties of Internet traffic), to realize the properties of packet fair scheduling with low complexity that is amenable to hardware implementation. In this context, tiered service implies that flows cannot request arbitrary weights (bandwidth shares), but must select among a small number of pre-determined weights (i.e., tiers). The directionality constraints in this case ensure that each flow subscribes to a tier that offers a share of the bandwidth at least as high as the flow requires.

- **Task Scheduling in Multiprocessor, Cluster, or Grid Systems.**
  Multiprocessor facilities, computational clusters, and Grid systems allow multiple users to share a pool of high-performance computational resources. If such a system provides only a small set of service levels, the scheduling, management, and handling of dynamic task requests can be simplified significantly. Consider, for instance the preemptive scheduling of a set of n periodic tasks on m identical processors [56, 73, 74]. A periodic task is made up of subtasks, each of length one, which have to be scheduled at regular intervals. In the discrete-time version of the problem, time is assumed to be slotted, and each task has a density $\rho_i, 0 < \rho_i < 1$, that represents the task's demand for processing time, in terms of subtasks per slot.

A feasible schedule for this problem exists if and only if $\sum_{i=1}^{n} \rho_i \leq m$, i.e., the total demand does not exceed the capacity of the m-processor system. A *proportionally fair* schedule [10] closely mimics the ideal fluid system in which both time and the subtasks are infinitely divisible. The fastest algorithm for constructing a proportionally fair schedule whenever one exists takes time $O(m \log n)$ *per slot* [5], hence it cannot scale to systems with a large number n of tasks (users). On the other hand, quantizing the task densities ρ_i to a small set of service tiers makes it possible to reduce the complexity of the proportionally fair algorithm to $O(m)$ [55], that is *independent* of the user population, and for a given system (i.e., a given number m of processors), it is constant.

Chapter 3
Bandwidth Tiered Service: Deterministic Demands

Consider a network operator providing a service characterized strictly by the amount of bandwidth allocated to each user, e.g., as is the case for many residential and business Internet access services. The service provider is interested in making available only a small number p of distinct bandwidth levels (tiers) to which users may subscribe. In offering such a tiered bandwidth service, the operator must determine the number of tiers and the amount of bandwidth for each tier, as well as provision sufficient capacity within the network. Assuming that the provider has some information regarding the users' bandwidth requirements, it must balance two objectives in dimensioning its network for the tiered service: on the one hand, it must ensure that each user's bandwidth needs are adequately satisfied, while on the other hand it must minimize its own cost for providing the service.

In this chapter we make the assumption that the service provider has complete knowledge of the exact bandwidth requirements of its customers; we address the problem of stochastic bandwidth demands in Chapter 5. We show that the problem of dimensioning the network for tiered bandwidth service with deterministic demands can be expressed as a directional p-median problem on the real line (DPM1). We address two variants of the problem, one in which the number p of tiers is provided as an input parameter, and one in which the number of tiers is a variable to be optimized, and we present efficient optimal algorithms for both. We also quantify the impact of this "bandwidth quantization" on the amount of network resources to be provisioned for the tiered service.

3.1 Bandwidth Tiered Service as a DPM1 Problem

Consider a packet-switched network serving n users. Let x_i be the amount of bandwidth requested by user i. The network offers $p \geq 1$ tiers of service; typically, $p \ll n$. The j-th level of service corresponds to bandwidth z_j, $z_1 < z_2 < \ldots < z_p$. In this section we assume that the number p of service tiers is fixed and known; we will

G.N. Rouskas, *Internet Tiered Services*, DOI: 10.1007/978-0-387-09738-1_3,
© Springer Science + Business Media, LLC 2009

remove this assumption in Chapter 3.4. For notational convenience, we let z_0 denote the "null" service level.

To ensure that the users' quality of service (QoS) requirements are met, the network provider maps each user i to service level z_j such that $z_{j-1} < x_i \leq z_j$. The additional bandwidth $z_j - x_i(\geq 0)$ allocated to user i represents the performance *penalty* associated with the tiered service. In other words, by mapping a user to the next higher offered service level, a tiered-service network may use more resources (bandwidth) than a continuous-rate one to satisfy the same set of requests for service. Since there is a cost associated with provisioning resources, it is important, from the point of view of the network provider, to minimize this additional bandwidth associated with the tiered service.

Specifically, given the set $X = \{x_i\}$ of user requests and the number of service levels p, we are interested in finding the set of service levels $Z = \{z_1, \ldots, z_p\}$ that minimizes the performance penalty over all n users; we refer to such set as "optimal." We now note that if we interpret bandwidth requests as "demand points" and service tiers as "supply points" (terminology standard in discrete location literature)[1], then the performance penalty of tiered service is expressed by the objective function (2.16). Consequently, the problem the network provider faces is equivalent to the directional p-median problem on the real line (Problem 2.4, DPM1) that we defined in Chapter 2.2.

It is well-known that the classical p-median problem on the real line (PM1) can be solved in polynomial time [46]. However, algorithms for the classical problem cannot be applied to solve the directional variant. Such algorithms have the option of assigning a demand point x to a supply point z that may lie either on the left (i.e., $z < x$) or the right (i.e., $x < z$) of it, depending on the distances between x and the supply points. In the former case, the resulting solution would violate the directionality constraints (2.15) and hence would not be feasible for DPM1.

We now develop an optimal dynamic programming algorithm [15] for the DPM1 problem; we present a more efficient algorithm in the next section. Both algorithms are based on the following two observations regarding the nature of the optimal supply points. First, note that for any feasible supply set for which $x_n < z_p$, the value of the objective function (2.16) can be reduced by setting $z_p = x_n$. Therefore, in an optimal supply set the maximum supply point z_p must equal the maximum demand point x_n. Furthermore, we can state the following lemma:

Lemma 3.1. *Let X be a set of n demand points such that $x_1 \leq x_2 \ldots \leq x_n$. There exists an optimal supply set $Z = \{z_1, \ldots, z_p\}$ such that $z_j \in X$, for each $j = 1, \ldots, p$.*

Proof. Suppose that there exists an optimal supply set of X called $Z' = \{z'_1, \ldots, z'_p\}$, $z'_1 < z'_2 < \ldots < z'_p$, for which there exists some $z'_a \in Z'$ but $z'_a \notin X$. Then, there can be no x_i that is mapped to z'_a. If there were, then z'_a could be moved down to $z'_a - \varepsilon$, for some $\varepsilon > 0$, lowering the objective and contradicting the optimality of Z'. Therefore, we can create Z from Z' by setting $z_j = z'_j$ for $j \neq a$, and $z_a = x_n$. $\square$

---

[1] We will use the terms "service (or bandwidth) demands," "service (or bandwidth) requests," and "demand points" interchangeably in this book; similarly for the terms "service (or bandwidth) tiers," "service (or bandwidth) levels," and "supply points."

Define $F(X,p)$ as the optimal value of the DPM1 objective function (2.16) when the demand set is X and the number of supply points is p. Also define $\Psi(X,p)$ as the bandwidth allocated to the users under to tiered service in the optimal solution for demand set X and number p of supply points. We can now rewrite the optimal value of the DPM1 objective function (2.16) as:

$$F(X,p) = \sum_{j=1}^{p}(n_j z_j) - \sum_{i=1}^{n} x_i = \Psi(X,p) - \sum_{i=1}^{n} x_i. \qquad (3.1)$$

In the above expression, the rightmost term represents the amount of bandwidth requested by the original set of demand points, i.e., the total amount of bandwidth that would be allocated to the users in a continuous-rate network, while $F(X,p)$ is the amount of excess bandwidth needed by the tiered-service network.

Based on the fact that for a given demand set X the rightmost term in expression (3.1) is constant, it is possible to solve the DPM1 problem by using the following dynamic programming algorithm to compute $\Psi(X,p)$ recursively, where $X_k = \bigcup_{i=1}^{k}\{x_i\}, k = 1,2,\ldots,n$, is the set with the k smallest demand points in $X(=X_n)$:

$$\Psi(X_1,l) = x_1, \quad l = 1,\ldots,p \qquad (3.2)$$
$$\Psi(X_k,1) = kx_k, \quad k = 1,\ldots,n \qquad (3.3)$$
$$\Psi(X_k,l+1) = \min_{q=1,\ldots,k-1} \{\Psi(X_q,l) + (k-q)x_k\}$$
$$l = 1,\ldots,p-1, k = 2,\ldots,n \qquad (3.4)$$

Expression (3.2) states that if there is only one demand point, it is the optimal supply point. Expression (3.3) is due to the fact that when $p=1$, the optimal supply point is equal to the largest demand point. The recursive expression (3.4) can be explained by noting that the $(l+1)$-th supply point must be equal to the demand point x_k. If the l-th supply point is equal to $x_q, q = 1,\ldots,k-1$, the tiered-service load is given by the expression in brackets in the right-hand side of (3.4), since $k-q$ demand points are mapped to supply point x_q. Taking the minimum over all values of q provides the optimal value.

The running time complexity of the above dynamic programming algorithm is $O(pn^2)$. Note that it is typical for network operators to serve a number of customers in the order of hundreds of thousands or more. Consequently, applying this algorithm of quadratic complexity to large problem instances can be challenging, especially for carrying out a "what-if" analysis by exploring a wide range of scenarios (e.g., in terms of the user demands). In the next section we show how to exploit a property of DPM1 to develop an optimal algorithm of linear complexity that scales well to large problem instances.

3.2 A Linear Complexity Algorithm for DPM1

To develop a more efficient algorithm for DPM1, we restate the problem as a constrained shortest path (CSP) problem on a directed acyclic graph (DAG) (Garey and Johnson's [39] problem ND30, see also [61]). We also show that the arc weights in the DAG representation of DPM1 obey the concave Monge property, allowing a solution in time $O(pn)$.

3.2.1 Graph Representation of DPM1

The following lemma establishes that solving DPM1 is equivalent to finding a minimum weight p-link path in a DAG.

Lemma 3.2. *Solving an instance of DPM1 with n demand points, demand set $X = \{x_i\}$, and p supply points is equivalent to finding a minimum weight p-link path from vertex 0 to vertex n in a weighted DAG with vertex set $V = \{0,1,\ldots,n\}$.*

Proof. Given an instance of DPM1, we construct a weighted, complete DAG as follows. Demand x_i gives rise to node i, and we create a dummy node 0. Arc weight $w(i,k)$ represents the cost of mapping demand points $i+1, i+2, \ldots, k$, to point k:

$$w(i,k) = \begin{cases} 0, & k = i+1 \\ (k-i-1)x_k - \sum_{j=i+1}^{k-1} x_j, & k > i+1 \end{cases}. \tag{3.5}$$

A path $t = (i_0,i_1),(i_1,i_2),\ldots,(i_{r-1},i_r)$ is a $(0\text{-}n)$-path if $i_0 = 0$ and $i_r = n$. The weight of path t is:

$$w(t) = w(i_0,i_1) + w(i_1,i_2) + \ldots + w(i_{r-1},i_r). \tag{3.6}$$

Any p-link path $t = (i_0,i_1),(i_1,i_2),\ldots,(i_{p-1},i_p)$ with $i_0 = 0$ and $i_p = n$ is a feasible solution for DPM1, with the following interpretation: the demand points corresponding to nodes $i_1,i_2,\ldots,i_p$, are designated as supply points and the weight of the path is the value of the corresponding objective function (2.16). Hence, finding the minimum weight path from vertex 0 to n that has exactly p arcs produces the optimal solution to the corresponding DPM1 instance. □

As an example, Fig. 3.1 shows the graph for a DPM1 instance with $n = 5$. Suppose $p = 3$, and let (0,2,4,5) be a 3-link path in Fig. 3.1. Then the corresponding feasible solution for DPM1 is $z_1 = x_2$, $z_2 = x_4$, and $z_3 = x_5$. The sum of the arc weights for this path equals the objective function value for the implied mapping for the corresponding solution $Z = \{z_1,z_2,z_3\}$, namely:

$$w(t) = w(0,2) + w(2,4) + w(4,5) = (x_2 - x_1) + (x_4 - x_3) + 0. \tag{3.7}$$

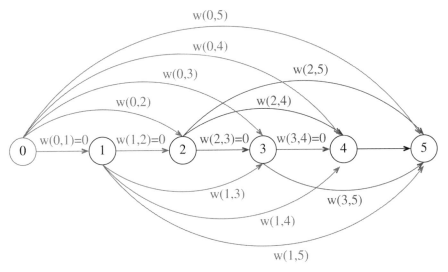

Fig. 3.1 DAG representation of an instance of the directional p-median problem with $n = 5$ (© 2007 IEEE)

3.2.2 Monge Condition and Totally Monotone Matrices

A weighted DAG satisfies the concave Monge condition if

$$w(i,j) + w(i+1,j+1) \leq w(i,j+1) + w(i+1,j) \tag{3.8}$$

holds for all $0 < i+1 < j < n$. We have the following result:

Lemma 3.3. *The arc weights of the graph representation of any DPM1 instance obey the concave Monge condition.*

Proof. Using the weights defined in (3.5), we evaluate the left hand side (LHS) and right hand side (RHS) of equation (3.8) as follows:

$$\begin{aligned}
\text{LHS} &= w(i,j) + w(i+1,j+1) \\
&= (j-i-1)x_j - \sum_{m=i+1}^{j-1} x_m + (j-i-1)x_{j+1} - \sum_{m=i+2}^{j} x_m.
\end{aligned} \tag{3.9}$$

$$\begin{aligned}
\text{RHS} &= w(i,j+1) + w(i+1,j) \\
&= (j-i)x_{j+1} - \sum_{m=i+1}^{j} x_m + (j-i-2)x_j - \sum_{m=i+2}^{j-1} x_m.
\end{aligned} \tag{3.10}$$

By subtracting (3.9) from (3.10) we can show that $\text{RHS} - \text{LHS} \geq 0$, hence that the concave Monge property (3.8) is satisfied.

$$\text{RHS} - \text{LHS} = (j - i - (j - i - 1))x_{j+1} + (j - i - 2 - (j - i - 1))x_j$$
$$+ \left(\sum_{m=i+1}^{j-1} x_m - \sum_{m=i+1}^{j} x_m + \sum_{m=i+2}^{j} x_m - \sum_{m=i+2}^{j-1} x_m \right)$$
$$= x_{j+1} - x_j + (-x_j + x_j)$$
$$= x_{j+1} - x_j$$
$$\geq 0. \tag{3.11}$$

The last step above follows from the fact that $x_1 \leq x_2 \leq \ldots \leq x_n$. $\square$

Consider a matrix M of real elements, and let $I(t)$ denote the index of the left-most column containing the maximum value in row t of M. Matrix M is said to be *monotone* if

$$t_1 > t_2 \ \Rightarrow \ I(t_1) \geq I(t_2), \quad \forall \, t_1, t_2. \tag{3.12}$$

Matrix M is said to be *totally monotone* if all its sub-matrices are monotone [2]. It has been shown [3] that a 2-dimensional Monge array is totally monotone.

An algorithm that can find the minimum entry in each column of a totally mono-tone $n \times m$ matrix, $n \geq m$, in $\Theta(n)$ time was developed in [2]. This elegant matrix searching algorithm has many geometric applications, and we show next how it can be applied to obtain a faster optimal algorithm for DPM1. For the details of the matrix searching algorithm, the reader is referred to [2, 9].

3.2.3 Efficient Dynamic Programming Algorithm for DPM1

The faster, linear complexity algorithm for DPM1 is based on the following dy-namic programming formulation to obtain the optimal value of the objective func-tion in (3.1):[2]

$$F(X_1, l) = 0, \quad l = 1, \cdots, p \tag{3.13}$$
$$F(X_k, 1) = w(0, k), \quad k = 1, \cdots, n \tag{3.14}$$
$$F(X_k, l+1) = \min_{q=l,\ldots,k-1} \{F(X_q, l) + w(q+1, k)\}$$
$$l = 1, \ldots, p-1, k = 2, \ldots, n \tag{3.15}$$

where $w(i,k)$ are the arc weights of the DAG corresponding to this instance of DPM1, as defined in (3.5). The optimal value for the objective function in (3.1) is obtained as the value of $F(X_n, p)$. Expressions (3.13)-(3.15) correspond to (3.2)-(3.4), respectively, and can be explained in a similar manner.

Our objective is to obtain the value of $F(X_n, p)$ by computing all the elements of each column l of the matrix defined by F in $O(n)$ time. Note that for $l = 1$, the elements of the first column can be computed in $O(n)$ time from expressions (3.14)

---

[2] In contrast, recall that the dynamic programming formulation (3.2)-(3.4) was used to compute the optimal value of the term $\Psi(X, p)$ in (3.1).

and (3.5). Therefore, we concentrate on computing expression (3.15) efficiently. To this end, we introduce a new function $\Gamma(q,k)$:

$$\Gamma(q,k) \;=\; F(X_q, l-1) + w(q+1, k), \quad q,k = 1,\ldots,n. \qquad (3.16)$$

We can now see that filling out the l-th column defined by matrix F, i.e., computing the n elements $F(X_k, l+1), k = 1, \cdots, n$, from expression (3.15) is equivalent to finding the minimum elements in each column of the $n \times n$ matrix defined by function $\Gamma(q,k)$. Also, $\Gamma(q,k)$ depends on the values of the elements of the $(l-1)$-th column of the matrix defined by F, which have already been calculated.

We now show that the function $\Gamma(q,k)$ obeys the concave Monge condition. From (3.8) we know that

$$w(q+1, k) + w(q+2, k+1) \le w(q+1, k+1) + w(q+2, k). \qquad (3.17)$$

If we add the term $F(X_q, l-1) + F(X_{q+1}, l-1)$ to both sides of (3.17) and use the definition of $\Gamma(q,k)$ in (3.16), we get:

$$\Gamma(q,k) + \Gamma(q+1, k+1) \le \Gamma(q, k+1) + \Gamma(q+1, j). \qquad (3.18)$$

Since $\Gamma(q,k)$ obeys the concave Monge condition, the matrix represented by $\Gamma(q,k)$ is totally monotone [3].

Based on the above observations, in order to solve the dynamic programming algorithm (3.13)-(3.15) we proceed by filling the $n \times p$ matrix defined by function $F(X_k, l)$ one column at a time. The first column ($l = 1$) is filled in $O(n)$ time using expression (3.14). In order to compute the n elements of the l-th column, $l = 2, \ldots, p$, we use expression (3.16) to form an $n \times n$ totally monotone matrix containing $\Gamma(q,k)$ values that depend on the values of the $(l-1)$-th column of the matrix F. The minimum elements in each column of the $n \times n$ matrix Γ are the n elements needed to fill the l-th column of matrix F. These elements can be obtained in $O(n)$ time using the algorithm in [2] we discussed in the previous section. Hence, the time to fill all p columns of matrix F, i.e., the time to find the optimal value for the objective function (3.1), is $O(np)$.

However, there remains one important issue that we need to address. The $O(n)$ algorithm in [2] assumes that the totally monotone matrix Γ is provided as input, since building this matrix would take time $O(n^2)$. In our case, the matrix Γ is not provided but has to be built anew for computing each column of matrix F. Rather than actually building the matrix in time $O(n^2)$, we now show how to evaluate the value of each element $\Gamma(q,k)$ in constant time. Whenever the algorithm in [2] needs to use the value of some element $\Gamma(q,k)$, rather than accessing the value from memory, we compute its value in constant time. By replacing one constant-time operation (memory access) with another (computing the value), we ensure that the algorithm runs in $O(n)$ time.

From (3.16) we see that element $\Gamma(q,k)$ depends on the values of $F(X_q, l-1)$ and $w(q+1, k)$. The value of $F(X_q, l-1)$ is already computed and hence can be accessed in constant time. In order to evaluate $w(q+1, k)$ in constant time, we per-

Table 3.1 Formulae for the PDF and CDF of the distributions of Fig. 3.2

Distribution	PDF	CDF	Domain
Uniform	1	x	$0 \leq x \leq 1$
Increasing	$2x$	x^2	$0 \leq x \leq 1$
Decreasing	$-2x+2$	$-x^2+2x$	$0 \leq x \leq 1$
Triangle	$4x$	$2x^2$	$0 \leq x < 0.5$
	$-4x+4$	$-2x^2+4x-1$	$0.5 \leq x \leq 1$
	$4/9$	$4x/9$	$0 \leq x < 0.25$
Unimodal	6	$6x-25/18$	$0.25 \leq x < 0.35$
	$4/9$	$4x/9+5/9$	$0.35 \leq x \leq 1$
	$1/4$	$x/4$	$0 \leq x < 0.25$
	4	$4x-15/16$	$0.25 \leq x < 0.35$
Bimodal	$1/4$	$x/4+3/8$	$0.35 \leq x < 0.65$
	1	$4x-33/16$	$0.65 \leq x < 0.75$
	$1/4$	$x/4+3/4$	$0.75 \leq x \leq 1$

form the following preprocessing operation before starting to solve the dynamic programming algorithm. Define, for $q = 1, \ldots, n$, $A(q) = \sum_{i=1}^{q} x_i$. It takes time $O(n)$ to compute and store the values $A(q)$ for all q. Once this is done, one can compute the value of $w(q,k)$ in constant time using the expression:

$$w(q,k) = -A(k) + A(q-1) + (k-q+1)x_k, \quad q \leq k. \qquad (3.19)$$

Hence the value of $\Gamma(q,k)$ can be calculated in constant time for any two values q and k.

3.3 Impact of Tiered Service on Network Resources

As we have already mentioned, the main drawback of tiered service is that it may require more resources than a service that provides each user with the exact amount of bandwidth requested. We now present simulation results to quantify this resource penalty as a function of the number p of tiers and the number n of users. To this end, we have generated problem instances with demand sets X drawn from six distributions: uniform, increasing, decreasing, triangle, unimodal, and bimodal, as shown in Fig. 3.2. The corresponding probability and cumulative distribution functions (PDF and CDF, respectively) are listed in Table 3.3. Note that in general, bandwidth demands will be in the range $(0, B]$, where B is the link capacity. However, in order to obtain results that are independent of the link capacity, we assume that all demands are normalized with respect to B; thus, the domain of the PDF and CDF of all distributions in Table 3.3 is $[0,1]$. From each input distribution, we generated 100 demand sets. Each demand set was generated starting from a unique seed for a Lehmer random number generator(RNG) [72,91] with modulus $2^{31} - 1$ and multiplier 48271.

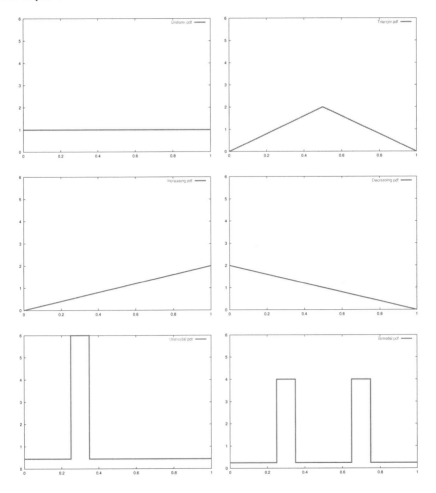

Fig. 3.2 Probability distribution functions (PDF) for the six demand distributions: uniform, triangle, increasing, decreasing, unimodal, and bimodal (© 2003 IEEE)

Returning to the objective function of the DPM1 problem in (2.16), we note that, for a given demand set X, the term $\sum_{j=1}^{p}(n_j z_j)$ represents the amount of bandwidth required by the tiered service with p tiers, whereas the term $\rho_X = \sum_{i=1}^{n} x_i$ is the minimum amount of bandwidth to satisfy all the demands in X. In order to be able to present aggregate results across different problem instances within a certain demand distribution, we define the following *normalized bandwidth requirement (NBR)* metric for each instance:

$$\text{Normalized bandwidth requirement (NBR)} = \frac{\sum_{j=1}^{p} n_j z_j}{\rho_X} \geq 1 \qquad (3.20)$$

Clearly, the closer this metric is to one, the lower the penalty in terms of excess resources needed due to tiered service. All the figures in this section plot the average value of this metric over 100 random problem instances from the stated demand distribution and values for the number n of users and p of service tiers.

Figs. 3.3-3.8 plot the value of the normalized bandwidth requirement metric corresponding to values to $p = 2, 3, \ldots, 100$, for the uniform, triangle, increasing, decreasing, unimodal, and bimodal distributions shown in Fig. 3.2, respectively. Three curves are shown in each graph, corresponding to demand sets of size $n = 100, 1000$ and 10000, respectively. Each point of a curve represents the average across 100 demand sets; we have also plotted 95% confidence intervals, but they are so narrow that they are hardly visible in the figures.

We observe that the general shape of the curves remains the same regardless of the input distribution: the normalized bandwidth requirement decreases sharply as p increases, and drops below 1.10 when $p \approx 15$. In fact, we have observed similar behavior for a wide range of distributions. Also, we note that the curve for $n = 100$ lies below the curve for $n = 1000$, which in turn lies below that for $n = 10000$. This result is consistent with intuition, since, for a fixed number p of service tiers, one would expect the "quantization penalty" to increase as the number of demands increases. However, the incremental penalty as we move from $n = 100$ to $n = 1000$ to $n = 10000$ diminishes, as evidenced by the fact that the curves for $n = 1000$ and $n = 10000$ almost coincide.

Figs. 3.9 and 3.10 plot the normalized bandwidth metric across all 100 problem instances generated from the triangle demand distribution; Fig. 3.9 shows results for problem instances with $n = 100$ demand points, while Fig. 3.10 shows results for $n = 1000$. Each figure contains curves for instances with $p = 2, 4, 6, 8, 10, 15$ and 20 supply points. Figures for the other five demand distributions exhibit similar characteristics, and are omitted.

The graphs in Figs. 3.9 and 3.10 present the information contained in Fig. 3.4, corresponding to the same triangle demand distribution, in a different way. A single point, say $p = 10$ in the $n = 100$ (respectively, $n = 1000$) curve of Fig. 3.4 is created by averaging the values of the 100 points shown in the curve for $p = 10$ in Fig. 3.9 (respectively, Fig. 3.10). As expected, we see that as the number p of supply points increases, the normalized bandwidth requirement improves; that is, as p increases, this metric approaches 1 from above. Comparing the curves in Fig. 3.9 to the corresponding ones in Fig. 3.10, we also note that the variation in the bandwidth requirement values across problem instances decreases as the demand set X increases in size from $n = 100$ to $n = 1000$.

Overall, the results shown in Figs. 3.3-3.10 indicate that by using 10-20 service tiers (supply points), a tiered-service network can adequately serve large user populations, dedicating no more than 5-8% bandwidth resources beyond the amount requested. Another interpretation is that, for a fixed amount of bandwidth, the tiered-service network can accept requests up to approximately 92-95% capacity. More importantly, our experiments indicate that this result is valid across a wide range of distributions of user requests, i.e., the bandwidth use of a tiered-service network is mostly insensitive to the actual demand distribution.

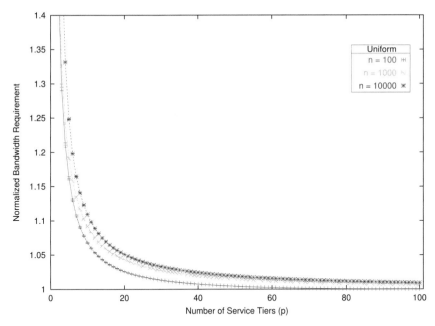

Fig. 3.3 Normalized bandwidth requirement vs. p, uniform distribution (© 2003 IEEE)

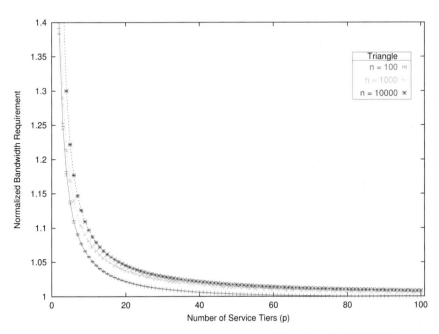

Fig. 3.4 Normalized bandwidth requirement vs. p, triangle distribution (© 2003 IEEE)

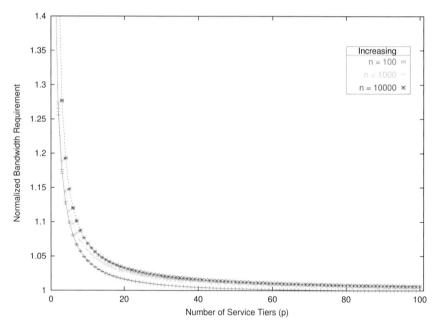

Fig. 3.5 Normalized bandwidth requirement vs. p, increasing distribution (© 2003 IEEE)

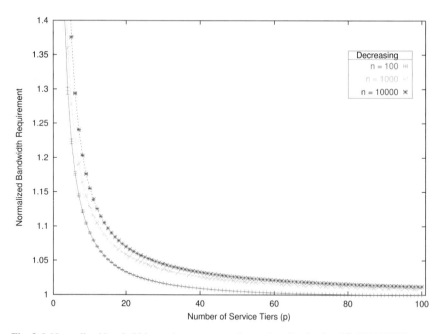

Fig. 3.6 Normalized bandwidth requirement vs. p, decreasing distribution (© 2003 IEEE)

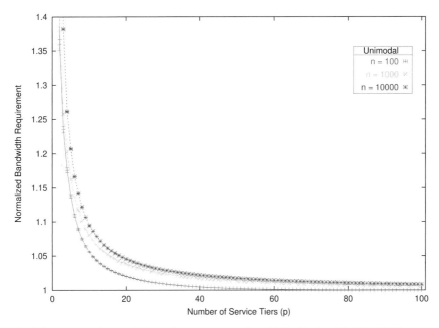

Fig. 3.7 Normalized bandwidth requirement vs. p, unimodal distribution (© 2003 IEEE)

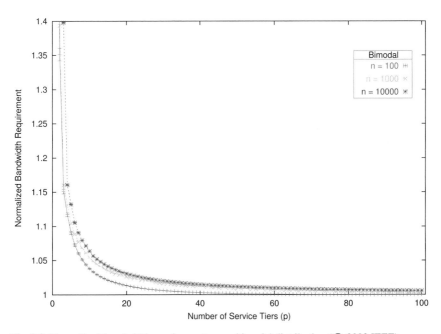

Fig. 3.8 Normalized bandwidth requirement vs. p, bimodal distribution (© 2003 IEEE)

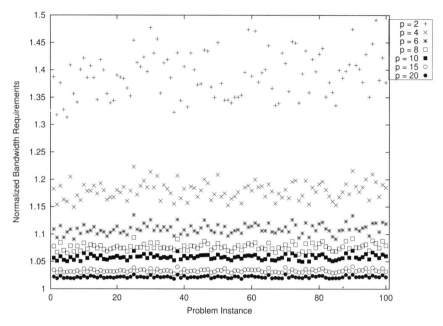

Fig. 3.9 Normalized bandwidth requirement for 100 problem instances, triangle distribution, $n =$ 100, various values of p (© 2003 IEEE)

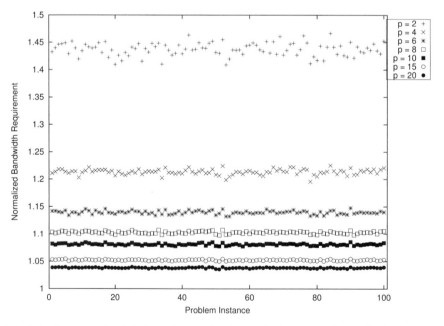

Fig. 3.10 Normalized bandwidth requirement for 100 problem instances, triangle distribution, $n =$ 1000, various values of p (© 2003 IEEE)

3.4 Joint Optimization of the Number and Magnitude of Service Tiers

In the definition of the DPM1 problem in Chapter 2.2, the number p of supply points is provided as an input parameter and there is no cost associated with a supply point. In a practical system, there may be a cost (e.g., additional hardware and/or software complexity) involved in providing each service tier. In this case, it is desirable to optimize *jointly* the number p of supply points and their placement, so as to strike a balance between the bandwidth penalty due to quantization and the cost of establishing additional supply points.

Let us assume that there is a cost associated with supporting a given number of service tiers, e.g., in terms of the relevant software and/or hardware mechanisms that would need to be implemented at the routers. Specifically, we assume that the *incremental* cost (e.g., due to additional queueing structures, policing mechanisms, control plane support, etc.) of offering one additional service tier is equal to α. Hence, the total cost for p tiers is αp.

The problem of jointly optimizing the number and magnitude of service tiers, a variant of DPM1 which we will refer to as *DPM1 with joint optimization (JDPM1)*, can be expressed as:

Problem 3.1 (JDPM1). Given a set X of n demand points, $x_1 \leq x_2 \leq \cdots \leq x_n$, find an integer $p \leq n$ and a feasible set Z of p supply points, $z_1 < z_2 < \cdots < z_p$, that minimizes the following objective function:

$$G(X,p) = \left(\sum_{j=1}^{p} (n_j z_j) - \sum_{i=1}^{n} x_i \right) + \alpha p \qquad (3.21)$$

where n_j is the number of demand points mapped to supply point s_j.

JDPM1 belongs to the family of uncapacitated facility location problems [85]. This family of problems include as part of the input a facility-build cost (in this case, a service tier cost) α, and the objective is to optimize jointly the number of facilities and their locations. Of course, due to the directionality constraint (i.e., the requirement that users subscribe to a service tier at least as large as their bandwidth demand), algorithms for existing location problems in which a demand may be served by any facility, to the left or right of it in the real line, cannot be applied to the JDPM1 problem.

We observe that the two terms in the right hand side of (3.21) behave differently as a function of $p = 1, \cdots, n$. Specifically, as p increases, the first term that represents the error due to quantization decreases, but the second term increases linearly with p. Hence, the optimal number of supply points will depend on the cost α. However, given a fixed number p of supply points, the optimal *placement* of these points is independent of the value of α (since for a fixed p the second term in (3.21) becomes constant).

Based on the above observations, one straightforward approach to tackling this joint optimization problem would be to run the linear dynamic programming algorithm for DPM1 in Chapter 3.2 n times, once for each value of $p = 1, \cdots, n$. Since the running time of the algorithm for a fixed value of p is $O(pn)$, the running time complexity of this approach is $O(n^3)$.

We now present a more efficient dynamic programming algorithm for JDPM1 by modifying the boundary conditions and the cost of a solution of the dynamic programming formulation (3.13)-(3.15) for DPM1, to account for the additional term in the objective function. Recall that X_k is the subset with the k smallest elements of the demand set $X = X_n$. Then, we can obtain the optimal value for the objective $G(X, k)$ using the following recursion:

$$G(X_1, l) = -\alpha l, \quad l = 1, \cdots, n \tag{3.22}$$
$$G(X_k, 1) = w(0, k) - \alpha, \quad k = 1, \cdots, n \tag{3.23}$$
$$G(X_k, l + 1) = \max_{q = l, \cdots, k-1} \{G(X_q, l) + w(q + 1, k)\} - \alpha$$
$$l = 1, \cdots, n - 1; \; k = 2, \cdots, n \tag{3.24}$$

Expressions (3.22)-(3.24) are similar to expressions (3.13)-(3.15), respectively. The main difference is the introduction of the service tier ("facility") cost α, which decreases the service provider surplus accordingly. For instance, in expression (3.24), the cost of the additional (i.e., $(l+1)$-th) service tier is accounted for in the right hand side by subtracting the value of parameter α.

At the end of the recursion, the entries of the last row of the table G, i.e., the values of $G(X_n, l), l = 1, \ldots, n$, correspond to the optimal value of the objective function for demand set X when there are l service tiers. Let $l^\star$ be the optimal value of l, i.e., a value such that $G(X_n, l^\star) \geq G(X_n, l)$ for all $l, l = 1, \cdots, n$. The value $l^\star$ and the corresponding set of supply points comprise the optimal solution to the JDPM1 problem.

It is straightforward to show that the $n \times n$ matrix G also satisfies the concave Monge condition as defined in Chapter 3.2.2. Consequently, the time complexity of the above dynamic programming algorithm is $O(n^2)$ using the implementation we described in Chapter 3.2.3. Finding the optimal value $l^\star$ by searching the last row of matrix G takes time $O(n)$, hence the overall time complexity of the algorithm is $O(n^2)$.

Fig. 3.11 plots the objective function (3.21) against the number p of service tiers for the uniform distribution (refer to Table 3.3) and $n = 1000$ demand points. Two curves are shown, each corresponding to a different value of the cost α ($= 0.1, 1$, respectively) for establishing an additional supply point. As we can see, the exact shape of the curves depends on the value of α, but both exhibit the same trend. Specifically, for $p = 1$, the objective function is equal to $nx_n + \alpha$, where x_n is the largest demand point; since the demand points are generated from a uniform dis-

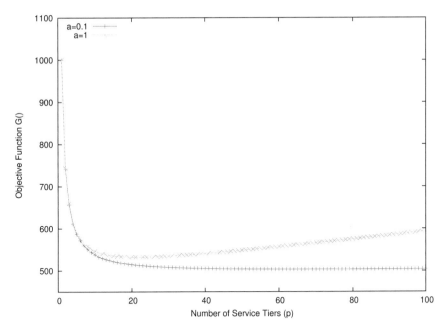

Fig. 3.11 Joint optimization of the number and placement of service tiers, uniform distribution, $n = 1000$, $\alpha = 0.1, 1$

tribution in $(0, 1)$, this value is approximately $n + \alpha$. For the selected values of α, the first term of the objective function dominates in the region where p is small. As p increases, initially the objective function decreases, reflecting a decrease in the dominant first term, until a minimum is reached; the optimal value of p at which the minimum is reached is different for each curve, as it depends on the value of α. Further increases of p beyond this optimal value result in an increase in the overall objective value, reflecting the fact that the second term αp becomes dominant for large p.

Chapter 4
Bandwidth Tiered Service: TDM Emulation

Many existing networks operate over an infrastructure that is based on time division multiplexing (TDM); examples include the legacy telephone network, the synchronous optical network/synchronous digital hierarchy (SONET/SDH) transport architecture, and time division multiple access wireless communications networks such as those employing the global system for mobile communications (GSM) standard. Whereas networks based on TDM technology today offer a multitude of data services in addition to classical voice service, it is often convenient for providers to allocate bandwidth to users in multiples of a unit rate that is typically tied to the slot size of the underlying transmission system. Legacy telephone networks and SONET/SDH networks, in particular, impose a rigid hierarchy of rates that are multiples of a voice channel (i.e., 64 Kbps), hence there is a proliferation of data services with rates ranging from DS-1 (1.5 Mbps) and DS-3 (45 Mbps) to STS-3 (155 Mbps) and STS-48 (2.5 Gbps) and beyond. These services are implemented by mapping a user's payload directly onto specific time slots of the transport system, e.g., using the generic framing procedure (GFP) [18, 49, 53].

In this chapter we introduce the concept of *TDM emulation* to describe a tiered-service network that allocates bandwidth in *arbitrary* multiples of a basic bandwidth unit (data rate). A packet-switched network operating with such a set of service levels would resemble a TDM network. Consequently, many robust network management functions developed for telecommunications networks, including admission control, routing, traffic engineering and grooming, etc., could be easily adapted for the tiered-service packet-switched network. The key premise of TDM emulation is that both the bandwidth unit and the service tiers be *configurable* (under software control) rather than fixed by the intrinsic structure of the underlying transport architecture as is the case with classical SONET/SDH networks. TDM emulation can be useful in a wide range of networking contexts and applications, including, but not limited to:

- **Bandwidth allocation in native packet-switched networks.** We emphasize that, as defined here, TDM emulation only affects the way that bandwidth is allocated to network users, *not* the data plane operation of packet-switched networks, e.g., those employing 1 or 10 Gigabit Ethernet (GE) links. For example, while

G.N. Rouskas, *Internet Tiered Services*, DOI: 10.1007/978-0-387-09738-1_4, bandwidth is allocated in multiples of the basic unit, users (flows) are not limited to using a particular slot. Similarly, unlike TDM networks where an unused slot is wasted, excess bandwidth can be allocated to active flows by the scheduling algorithm. Furthermore, the bandwidth unit is not fixed or determined by hardware, as in a TDM network, but, it is configurable and can be optimized for the characteristics of the carried traffic. In addition, the routers provide for free the functionality of a time-slot interchange. Overall, with TDM emulation, tiered-service packet-switched networks may enjoy many of the benefits, in terms of control and management, of a TDM network, but without the data plane rigidities of such a network.

- **Flexible bandwidth allocation in next generation SONET/SDH networks.** The emergence of framer technologies such as virtual concatenation (VCAT) [21] coupled with the link capacity-adjustment scheme (LCAS) [54] enable providers to make significantly more efficient use of the existing SONET/SDH infrastructure [22,33,48,54]. Virtual concatenation allows finer granularity for provisioning of bandwidth services, making it possible to configure service tiers in any multiple of 64 Kbps[1]. LCAS, on the other hand, offers a means to reconfigure (i.e., enlarge or shrink) dynamically the size of a SONET/SDH data pipe without impacting the transported data, permitting users to move between service tiers as their requirements evolve.

- **Traffic grooming.** Traffic grooming [34] refers to techniques used to combine low-speed traffic streams for transport over high-speed wavelengths so as to minimize the network cost in terms of line terminating equipment and/or electronic switching [35,79,92,121]. The traffic grooming problem has been investigated extensively in the literature [24, 40, 113, 120], but most studies make the assumption that traffic demands are multiple of a basic unit, e.g., STS-3 for an OC-48 link; this assumption is a reflection of the dominance of SONET/SDH in the transport network infrastructure. Packet networks employing TDM emulation (e.g., MPLS networks with LSP sizes that are multiples of a basic unit) may employ these traffic grooming solutions without any modifications, and thus leverage the vast body of research that already exists.

4.1 TDM Emulation As A Constrained DPM1 Problem

In a tiered-service network with TDM emulation, all service tiers are multiples of some quantity r that represents the unit bandwidth, i.e., the smallest amount in which bandwidth may be allocated. In packet switched networks, r is itself a configurable parameter, and is not tied to any underlying, hardware-imposed slot structure. Therefore, the objective in such a network is to select *jointly* the unit rate r and the service tiers in some optimal manner. To this end, in this chapter we pose the optimization

---

[1] Although in this case the basic rate is determined by the SONET/SDH hardware and hence fixed, we will see shortly that 64 Kbps is a reasonable unit for allocating bandwidth due to the overhead involved in making the bandwidth unit too small.

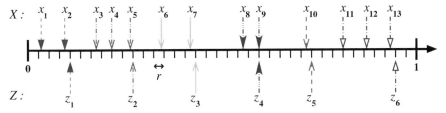

Fig. 4.1 Sample mapping of demand points x_i to supply points z_j that are multiples of a basic unit of magnitude r (© 2007 IEEE)

problem arising in TDM emulated networks as a DPM1 problem with an additional constraint on the values of the optimal supply points.

Fig. 4.1 illustrates the relationship of TDM emulation to DPM1 by showing a sample mapping from a set X of 13 demand points onto a set S of 6 supply points, under this additional constraint. The set X is identical to the one in Fig. 2.2 which shows a sample mapping under the original DPM1 problem. The main difference is that with the new constraint, the supply points are all multiples of the same unit r.

We will refer to the variant of the directional p-median problem with the constraint that all supply points are multiples of the unit bandwidth r as TDM-DPM1. Recalling that n_j denotes the number of demand points mapped to supply point z_j, TDM-DPM1 can be expressed as:

Problem 4.1 (TDM-DPM1). Given a set X of n demand points, $x_1 \leq x_2 \leq \ldots \leq x_n$, and a constant β, find a real r and a feasible set Z of p supply points, $z_1 < z_2 < \cdots < z_p$, $1 \leq p \leq n$, so as to minimize the objective function:

$$H(X,p) = \left(\sum_{j=1}^{p} (n_j z_j) - \sum_{i=1}^{n} x_i \right) + \frac{\beta}{r} \qquad (4.1)$$

under the constraints

$$z_j = r k_j, \quad k_j : \text{integer}, \quad j = 1, \ldots, p. \qquad (4.2)$$

The first term in the right hand side of (4.1) is the objective function (2.16) of the DPM1 problem and represents the excess bandwidth penalty, as before. In our formulation, the objective function for TDM-DPM1 includes the additional term $\frac{\beta}{r}$, where β is some constant related to the operation of the system, as we explain shortly. The presence of a term which is a monotonically decreasing function of r in (4.1) is necessary, since without it TDM-DPM1 reduces to DPM1: if nothing prevents r from being very small, then the optimal is obtained for $r = 1$ bps as the solution to DPM1 which minimizes the excess bandwidth penalty.

More importantly, the term $\frac{\beta}{r}$ is of practical significance as it captures the overhead associated with making the unit r of bandwidth allocation small. To illustrate, let us make the simplifying assumption that all users request and receive the basic rate of r bits/sec. After serving a user, the system incurs some overhead due to the bookkeeping operations, memory lookups, etc., required before it can switch to serving another user. Let τ denote the amount of time required to switch between users, *expressed as the number of bits that could be transmitted during this time at the given service rate*. Therefore, the quantity $\frac{\tau}{r}$ represents the amount of overhead operations relative to the bandwidth unit. This relative overhead, which increases as the unit of bandwidth decreases, is similar in principle to the "cell tax" incurred in carrying IP traffic over ATM networks due to the relatively large fraction of header (i.e., overhead) bits to data bits. In the objective function (4.1) we use the term $\frac{\beta}{r}$ where $\beta = c\tau$ and c is a constant which ensures that the two terms in the rightmost side of (4.1) are expressed in the same units.

We note that the first term in the right hand side of (4.1) requires that the unit r be small so as to minimize the excess bandwidth. However, making r small would increase the second term $\frac{\beta}{r}$ which represents the bandwidth wasted due to overhead operations. Therefore, the objective of TDM-DPM1 is to determine the value of r so as to strike a balance between these two conflicting objectives.

4.1.1 Optimal Solution to TDM-DPM1 for Fixed r

As defined, the objective of TDM-DPM1 is to find jointly optimal values for the basic bandwidth unit r (a real number) and the p supply points. However, let us consider for a moment the special case where the value of r is fixed and not subject to optimization; as we shall see shortly, the algorithm for this problem is useful in tackling the general one. In this case, the term $\frac{\beta}{r}$ in (4.1) is constant and does not affect the minimization. Hence, the objective function is identical to that of the DPM1 problem.

Consider an instance of TDM-DPM1 in which the value of the basic bandwidth unit is fixed at $r = r_0$; that is, the p supply points can only take the values $kr_0, k = 1, \ldots, K$. Let $U = \{u_1, \ldots, u_K\}$ be the set of candidate values for the p supply points, $u_k = kr_0$; in Fig. 4.1, these candidate values are represented by the ticks below the horizontal line. Integer K corresponds to the largest possible multiple of r_0, i.e., $K = \left\lceil \frac{x_n}{r_0} \right\rceil$, where x_n is the largest demand point.

This version of TDM-DPM1 can be represented by a DAG similar to the one in Fig. 3.1. In this case, the DAG has $K + 1$, rather than $n + 1$, vertices: the dummy vertex 0 and the K vertices corresponding to the K candidate values for the supply points (recall that in DPM1, the candidate supply points are the n demand points). Similarly, the arc weight $w(i, k)$ in this DAG represents the cost of mapping the demand points with values between candidate supply points u_i and u_k to u_k:

$$w(i,k) \quad = \quad \sum_{u_i < x_j \leq u_k} (u_k - x_j) \tag{4.3}$$

It is also not difficult to verify that these weights satisfy the concave Monge condition (3.8) for all $0 < i + 1 < j < K$.

Since this version of TDM-DPM1 has the same objective function as DPM1 and can be represented by a DAG whose weights satisfy the concave Monge condition, we can solve it optimally using the dynamic programming algorithm in Chapter 3.2. Note that the algorithm will run in $O(pK)$, not $O(pn)$, time, as it has to consider K candidates for the p supply points.

4.1.2 The Behavior Of The TDM-DPM1 Objective Function

To obtain insight into how the additional parameter r affects the optimization, let us investigate the behavior of the objective function $H(X,p)$ in (4.1) as we vary r. In Fig. 4.2 we plot the objective function against the value of r for an instance of TDM-DPM1 with $n = 1000$ demand points, $p = 10$ supply points, and $\beta = 0.01$, with the set X of n demand points generated from a uniform distribution in $(0,1)$ (refer also to Table 3.3). We varied the value of the basic unit r in increments of $\delta_r = 10^{-5}$ across the range shown in the figure. For each (fixed) value of r we obtained the optimal supply points in the manner we described in the previous subsection, from which we evaluated the objective function (4.1), including the term $\frac{\beta}{r}$.

The behavior exhibited in Fig. 4.2 is representative of the TDM-DPM1 instances that have been studied. At low values of r, the term $\frac{\beta}{r}$ representing the overhead cost dominates, resulting in large overall values. As r increases, there is an initial period of rapid decrease in the objective function as the term representing the excess bandwidth penalty starts to become important. Following this initial decrease, the curve settles into a seesaw pattern. The high and low points along this pattern depend on the values of the multiples of r relative to the demand points: when multiples of r are aligned close to demand points, there is little bandwidth penalty for mapping these demand points to supply points that are multiples of r, hence the objective function has a lower value; the opposite is true when there is a mismatch between multiples of r and demand points. We also note that as the value of r increases further, the curve trends upwards. This behavior is due to two factors that come into play when r becomes large: the excess bandwidth term in (4.1) starts to dominate, and at the same time this term increases in value as large values of r are too coarse to minimize the excess bandwidth.

It is clear from Fig. 4.2 that the TDM-DPM1 objective function is not convex and includes several troughs at irregular intervals. This non-convex nature makes standard optimization techniques (e.g., steepest descent methods [13]) impractical, as they are likely to get trapped in a local minima. Next, we describe an exhaustive search approach for identifying the value of r and the set of p supply points that

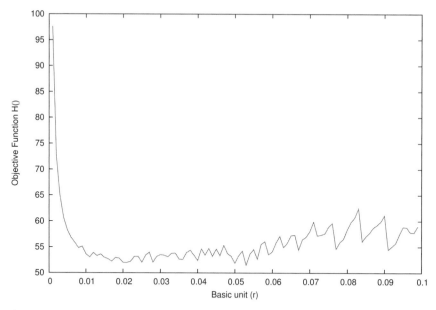

Fig. 4.2 The value of the objective function $H(X,p)$ against r, $n = 1000$, $p = 10$, $\beta = 0.01$, demand points generated from the uniform distribution in $(0,1)$ (© 2007 IEEE)

minimize the objective function $H(X,p)$ in (4.1), and in the following section we develop a suite of heuristics that trade solution quality for running time.

4.1.3 An Exhaustive Search Algorithm for TDM-DPM1

We first observe that for a TDM-DPM1 instance with p supply points and x_n the largest demand point, the largest value that r may take under the constraint that all p supply points be an integer multiple of r is $r_{max} = \frac{x_n}{p}$. Hence, the optimal value of r lies in the interval $(0, r_{max}]$. Let δ_r be a small increment value, and consider the set

$$R = \left\{ r_m = m\delta_r \leq r_{max} = \frac{x_n}{p}, \; m = 1, 2, \ldots, \left\lfloor \frac{r_{max}}{\delta_r} \right\rfloor \right\}. \tag{4.4}$$

The exhaustive algorithm selects the optimal value $r^\star$ among the ones in set R, and the corresponding optimal supply set $Z^\star$ using the steps in Fig. 4.3.

In order to determine the running time complexity of this exhaustive search algorithm, let L be the size of set R (i.e., the number of candidate values of r to be considered),

$$L = \left\lfloor \frac{r_{max}}{\delta_r} \right\rfloor = \left\lfloor \frac{x_n}{p\delta_r} \right\rfloor \tag{4.5}$$

Exhaustive Search Algorithm
Input: A demand set X, the number of supply points p, and a constant β.
User-defined parameter: An increment δ_r.
Output: A unit bandwidth r and a set Z of supply points.

begin
1. $R \leftarrow \left\{ r_m = m\delta_r \leq r_{max} = \frac{x_n}{p}, m = 1, 2, \ldots, \left\lfloor \frac{r_{max}}{\delta_r} \right\rfloor = L \right\}$
2. $r^\star \leftarrow r_1; m \leftarrow 1; H^\star \leftarrow \infty$
3. **while** $r_m < r_{max}$ **do**
4. $r \leftarrow r_m$ // fix the value of the unit rate
5.   Determine the set of candidate supply points
    $U_m \leftarrow \left\{ u_k = kr_m, k = 1, \ldots, K_m = \frac{x_n}{r_m} \right\}$
6.   Use the approach described in Chapter 4.1.1 to obtain the optimal set Z_m
    of p supply points for the fixed value r_m
7. $H_m \leftarrow$ the value of the objective function (4.1) for the set Z_m and $r = r_m$
8.   **if** $H_m < H^\star$ **then**
9. $r^\star \leftarrow r_m; H^\star \leftarrow H_m; Z^\star \leftarrow Z_m$
10.   **end if**
11. **end while**
12. **return** the optimal solution $r^\star$ and $Z^\star$, and the corresponding value of the
    objective function $H^\star$
end

Fig. 4.3 The exhaustive search algorithm for the TDM-DPM1 problem

and K_m be the number of candidate supply points when the value of $r = r_m$:

$$K_m = \frac{x_n}{r_m}. \tag{4.6}$$

The dynamic programming algorithm will be run L times, and during the m-th iteration, i.e., when $r = r_m$, the algorithm will take $O(pK_m)$ time. Since

$$\sum_{m=1}^{L} pK_m = \frac{px_n}{\delta_r} \sum_{m=1}^{L} \frac{1}{m} = \frac{px_n}{\delta_r}(\ln L + \gamma) \tag{4.7}$$

where $\gamma = 0.577\ldots$, is Euler's constant, the running time complexity of the algorithm is $O\left(\frac{px_n}{\delta_r} \ln\left(\frac{x_n}{p\delta_r}\right)\right)$.

As we can see, the complexity of the exhaustive search depends critically on the value of the increment δ_r which determines the granularity of the search. With finer granularity (i.e., smaller δ_r), the accuracy of the algorithm increases, but its complexity also increases dramatically; the opposite is true when δ_r becomes larger and the granularity coarser. We also note that the time complexity is independent of the number n of demand points, and depends only on the largest demand x_n. In the applications we consider, the input value x_n is bounded above by the bandwidth available on the highest capacity link in the network. To get a sense of the values involved in expression (4.7), consider a network with 10 Gbps links. A reasonable value for the bandwidth increment is $\delta_r = 64$ Kbps. Assuming that the largest demand can be equal to the capacity of a link, we have that

$$\frac{x_n}{\delta_r} \approx 10^6 \tag{4.8}$$

which demonstrates that the exhaustive search is taxing in terms of both computational and memory requirements.

4.1.4 Optimization Heuristics

We now present a set of heuristics for the TDM-DPM1 problem. Each heuristic trades solution quality for speed by using its own approach to reduce the size of the space of candidate values for r and/or the supply points that it considers.

4.1.4.1 Demand Driven Heuristic (DDH)

Recall that, for each candidate value r_m of r, the exhaustive search algorithm considers all the K_m multiples of r_m as the set of potential supply points, where K_m, given in expression (4.6) can be much larger than the number n of demand points. The intuition behind this heuristic is that the optimal supply points are more likely to be located just above a demand point, since otherwise there would be a larger penalty in terms of excess bandwidth. Specifically, the heuristic operates identically to the exhaustive search algorithm described in Fig. 4.3, with the only difference that, for a fixed value r_m, it considers only the n multiples of r_m that are located immediately to the right of (or coincide with) the n demand points. In other words, the set U of candidate values for the p supply points is

$$U = \left\{ u_i = r_m \times \left\lceil \frac{x_i}{r_m} \right\rceil, i = 1, \ldots, n \right\} \tag{4.9}$$

instead of the one in Step 5 of Fig. 4.3. Since there have to be n different candidate supply points, the range of values for r is in the interval $\left(0, \frac{x_n}{n} = r_{max}\right]$. Using n instead of K_m in expression (4.7) and the new value for r_{max} in expression (4.5), we find that the running time complexity of the DDH heuristic is $O\left(\frac{px_n}{\delta_r}\right)$, which represents an improvement over the exhaustive algorithm, especially for small values of δ_r which allow for a finer granularity search.

4.1.4.2 Supply Driven Heuristics

Both the DDH and the exhaustive search algorithms apply the dynamic programming algorithm in Chapter 3.2 for each candidate value for parameter r. We now present two heuristics based on the assumption that the optimal supply points for TDM-DPM1 are likely to be close to the optimal supply points for the corresponding unconstrained DPM1 problem with the same demand set. Therefore, each heuristic initially runs the dynamic programming algorithm for the corresponding DPM1 problem (refer to Chapter 3.2.3), and computes the optimal set

$$Z^{DPM1} = \{z_1^{DPM1}, \ldots, z_p^{DPM1}\} \qquad (4.10)$$

of supply points for that problem. This step takes time $O(pn)$, and this dynamic programming algorithm is not used again by the heuristics.

The first algorithm, which we call the *unidirectional supply driven heuristic (USDH)*, sets the i-th supply point for a given candidate value r_m of r to the smallest multiple of r_m that is greater than or equal to supply point z_i^{DPM1}. In other words, the set Z_m of supply points for candidate r_m is defined as

$$Z_m = \left\{ \left\lceil \frac{z_i^{DPM1}}{r_m} \right\rceil r_m, i = 1, \ldots, p \right\}. \qquad (4.11)$$

The heuristic returns the value r_m and corresponding set Z_m which result in the minimum value for the objective function (4.1). The operation of the heuristic is summarized in Fig. 4.4. Note that Step 1 of the algorithm takes time $O(pn)$, and the **while** loop between Steps 3 and 9 of the heuristic is executed $L = \left\lceil \frac{x_n}{p\delta_r} \right\rceil$ times. Each execution of the **while** loop takes time $O(p)$ to compute the supply points (in Step 4) and the objective function value (in Step 5). Therefore, the running time complexity of USDH is $O\left(pn + \frac{x_n}{\delta_r}\right)$.

The second algorithm is called the *bidirectional supply driven heuristic (BSDH)*, and computes a set of $2p$ possible values for the supply points for each candidate value r_m. The first set of p values is identical to the set Z_m in (4.11) used by the USDH algorithm above. In addition, this heuristic considers the set Z'_m consisting of the p largest multiples of r_m that are less than the corresponding supply points z_i^{DPM1}, i.e.,

$$Z'_m = \left\{ \left\lfloor \frac{z_i^{DPM1}}{r_m} \right\rfloor r_m, i = 1, \ldots, p \right\}. \qquad (4.12)$$

The $2p$ elements of these two sets collectively become the candidates for being one of the p supply points when the value of $r = r_m$. Consequently, the operation of the BSDH algorithm is identical to that of USDH in Fig. 4.4, with the exception that Step 4 is replaced by a call to the dynamic programming algorithm in Chapter 4.1.1 to select the optimal set of p supply points from the set $Z_m \bigcup Z'_m$; it is easy to see that the dynamic programming algorithm works even when the set of the candidate supply points is a proper subset of the set of all integer multiples of r_m. As with

Unidirectional Supply Driven Heuristic (USDH) Algorithm
Input: A demand set X, the number of supply points p, and a constant β.
User-defined parameter: An increment δ_r.
Output: A unit bandwidth r and a set Z of supply points.

begin
1. Compute the optimal set Z^{DPM1} of supply points under DPM1
1. $R \leftarrow \left\{ r_m = m\delta_r \leq r_{max} = \frac{x_n}{p}, m = 1, 2, \ldots, \left\lceil \frac{r_{max}}{\delta_r} \right\rceil = L \right\}$
2. $r^\star \leftarrow r_1; m \leftarrow 1; H^\star \leftarrow \infty$
3. **while** $r_m < r_{max}$ **do**
4. $Z_m \leftarrow$ the set of supply points from expression (4.11)
5. $H_m \leftarrow$ the value of the objective function (4.1) for the set Z_m and $r = r_m$
6.   **if** $H_m < H^\star$ **then**
7. $r^\star \leftarrow r_m; H^\star \leftarrow H_m; Z^\star \leftarrow Z_m$
8.   **end if**
9. **end while**
10. **return** the optimal solution $r^\star$ and $Z^\star$, and the corresponding value of the
    objective function $H^\star$
end

Fig. 4.4 The USDH algorithm for the TDM-DPM1 problem

USDH, the heuristic returns the value r_m and corresponding p supply points that minimize (4.1).

We expect the BSDH algorithm to perform better than USDH since it considers a larger number of candidate supply points. This improved performance, however, is at the expense of having to run the dynamic programming algorithm on a set of $2p$ points, meaning that Step 4 takes time $O(p^2)$ instead of $O(p)$. Hence, the running time complexity of BSDH is $O\left(p \left[n + \frac{x_n}{\delta_r} \right] \right)$.

4.1.4.3 The Power of Two Heuristic (PTH)

This heuristic simply selects the set of p supply points as the set of the p consecutive powers of two such that the largest element in the set is the smallest power of two that is larger than or equal to the largest demand point x_n; in other words,

$$ Z = \left\{ 2^{q+1}, 2^{q+2}, \ldots, 2^{q+p} \mid 2^{q+p-1} < x_n \leq 2^{q+p} \right\}. \tag{4.13} $$

In essence, the power of two heuristic (PTH) implements an *exponential tiering structure* that is similar to the ones currently being used by several DSL or cable ISPs [50, 51] (refer also to Chapter 1.2). It is also similar in spirit to approaches used in packet fair schedulers that assign packet flows in classes (e.g., as in [93]) whose boundaries are defined by powers of two.

The PTH solution is consistent with TDM-DPM1 in that it consists of supply points all of which are a multiple of a basic unit, in this case 2^{q+1}. However, as we shall see shortly, the excess bandwidth penalty for this solution can be quite high compared to the other algorithms. Therefore, we consider this solution here as a baseline case only.

4.2 Performance Evaluation

Let us now turn our attention to evaluating the performance of TDM emulation. Specifically, we present simulation results to (1) investigate the relative performance of the various algorithms for the TDM-DPM1 problem, and (2) determine the impact of TDM emulation on the operation of a network, compared to either a continuous-rate network or one with service tiers that are not subject to the constraint that they be multiples of the same unit. We consider two metrics in quantifying the impact of TDM emulation. The first measures the amount of resources (bandwidth) required to satisfy the user demands, and reflects the bandwidth penalty imposed on the network provider due to TDM emulation. The second metric captures the impact of TDM emulation on users in terms of the probability that their requests for service will be blocked in a dynamic network environment.

The demand sets X of the problem instances we consider throughout this section were generated from one of six distributions whose PDF and CDF are listed in Table 3.3. As in Chapter 3.3, we make the assumption that all demands are normalized with respect to the link capacity B; thus, the domain of all distributions is $[0,1]$, as shown in Table 3.3. Also, based on our discussion in Chapter 4.1.2 and the fact that the largest demand point $x_n \leq 1$, we used an increment value $\delta_r = 10^{-5}$ whenever applicable.

4.2.1 Algorithm Comparison

Let us first investigate the relative performance of the various algorithms for TDM-DPM1 with respect to the objective function (4.1). Fig. 4.5 plots the value of the objective function against the value of r for each of the four TDM-DPM1 heuristic algorithms we presented in this chapter: DDH, BSDH, USDH, and PTH. We also show the optimal value for the corresponding DPM1 problem; this value does not include the overhead term $\frac{\beta}{r}$ of (4.1), hence it serves as a lower bound for the TDM-DPM1 algorithms. These results were obtained for a problem instance with $n = 1000$ demand points generated from the triangle distribution of Table 3.3, $p = 15$ supply points, and overhead parameter $\beta = 0.05$. Note that the DPM1 and PTH solutions do not take parameter r into account, hence they are shown as horizontal lines in the figure.

From the figure we observe that the three algorithms DDH, BSDH, and USDH are able to obtain solutions that are close to the lower bound (i.e., the optimal value for the corresponding DPM1 problem) whenever the basic unit r is not very small.

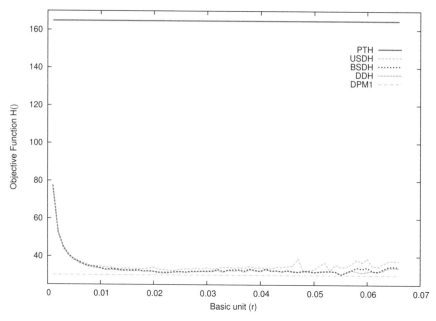

Fig. 4.5 Objective function value against r, $n = 1000$, $p = 15$, $\beta = 0.05$, triangle distribution (© 2007 IEEE)

On the other hand, as r decreases below 0.01, the curves for DDH, BSDH, and USDH increase rapidly. This behavior is inherent to the nature of the TDM-DPM1 objective function: at low values of r, the overhead term $\frac{\beta}{r}$ dominates, hence the optimal value for the TDM-DPM1 problem is expected to be significantly higher than the one for the corresponding DPM1 problem.

An important observation from this figure is that the solution produced by PTH performs much worse than the ones produced by the other algorithms. This result confirms our earlier observation that using powers of two to define classes of traffic is not an efficient approach, and indicates that the exponential tiering structures used by ISPs [50, 51] may be far from optimal. Since the results shown in Fig. 4.5 are representative of the behavior of PTH we have observed for all other problem instances and distributions, we do not consider this heuristic in the remainder of this chapter.

For the results shown in Fig. 4.6 and 4.7, we have generated thirty problem instances with $n = 100$, $p = 5$, and $\beta = 0.05$, from the increasing and triangle distributions, respectively. The figures plot the objective function value returned by each of four algorithms, DDH, BSDH, USDH, and DPM1, for each problem instance; again, the DPM1 solution provides a lower bound for the other three algorithms. The graphs show that, except for a few instances, all three TDM-DPM1 algorithms are close to the lower bound. Of the three algorithms, DDH produces the lowest objective function values, followed closely by BSDH. The objective function values returned by USDH are generally higher than those of the DDH and BSDH heuristics, but USDH has a much faster running time. Hence, these results indicate that there is a tradeoff between quality of solution and running time complexity of the algorithms.

Finally, Figs. 4.8 and 4.9 plot the objective function (4.1) against the basic unit r for solutions obtained by the DDH, BSDH, and USDH algorithms. Again we observe that the performance of the DDH and BSDH algorithms is very similar across the range of values of r shown here, while the faster USDH algorithm produces solutions that are no more than 20% above those of the other two algorithms.

4.2.2 Impact on the Network Provider: Bandwidth Penalty Due to TDM Emulation

Let us now turn our attention to determining the penalty in terms of excess resources needed due to tiered service. Given a demand set X, a continuous-rate link will use an amount of bandwidth equal to $\sum_i x_i$ to satisfy all the demands in X. A link of a tiered-service network, on the other hand, will in general use more bandwidth, as each demand x_i will be mapped to the next offered level of service (i.e., supply point). As in Chapter 3.3, we use the *normalized bandwidth requirement* metric, defined in equation (3.20) to characterize the bandwidth penalty incurred by the operator of a tiered-service network. As the reader may recall, the normalized bandwidth requirement is taken as the ratio of the amount of bandwidth $(=\sum_{j=1}^{p}(n_j z_j))$ used by a tiered-service network to the amount of bandwidth used by a continuous network; hence, it is desirable for this ratio to be close to one. Note also that this metric does not take intro account the overhead term $\frac{\beta}{r}$ in the objective function of the TDM-DPM1 problem, as it only reflects the bandwidth required to map users to the appropriate service tiers.

Figs. 4.10 and 4.11 plot this metric against the number p of service levels offered by the network. Each point in these curves is the average over 30 different problem instances generated by a uniform distribution; similar results have been obtained for all other distributions shown in Table 3.3 [9]. In both figures, the value of the overhead parameter $\beta = 0.05$.

Fig. 4.10 presents results for two tiered service scenarios: one in which the service levels are obtained from the DPM1 algorithm, and one in which they are obtained from the DDH algorithm; DDH is selected as a representative algorithm for the TDM-DPM1 problem in which the service levels are all multiples of a basic bandwidth unit r. As we can see, the curve for DDH is above the one for DPM1. This result is expected, since the (optimal) DPM1 algorithm is only concerned with minimizing the excess bandwidth due to tiered service, while the DDH algorithm also has to take into account the constraint that all service levels be multiples of a basic unit. However, the additional penalty due to the constraint imposed by the TDM-DPM1 problem is relatively small. Also, the normalized bandwidth requirement decreases rapidly with the number of service levels. This behavior was also

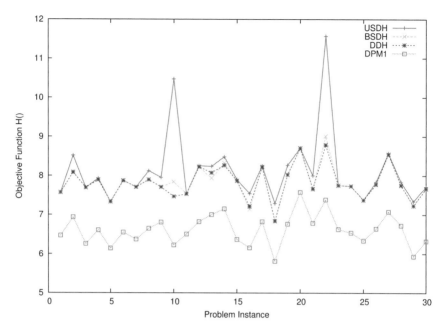

Fig. 4.6 Objective function value for each problem instance, $n = 100$, $p = 5$, $\beta = 0.05$, increasing distribution (© 2007 IEEE)

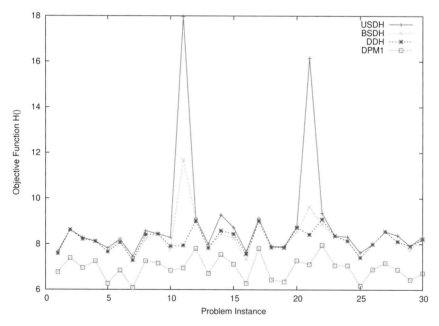

Fig. 4.7 Objective function value for each problem instance, $n = 100$, $p = 5$, $\beta = 0.05$, triangle distribution (© 2007 IEEE)

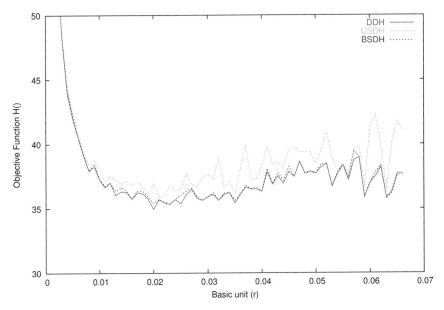

Fig. 4.8 Objective function value against r, $n = 100$, $p = 15$, $\beta = 0.05$, uniform distribution

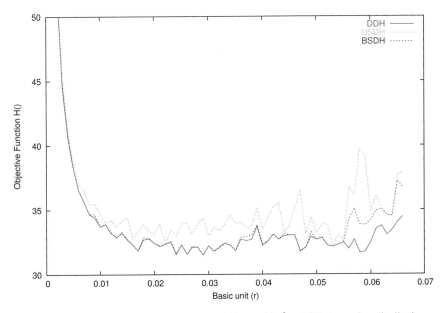

Fig. 4.9 Objective function value against r, $n = 100$, $p = 15$, $\beta = 0.05$, decreasing distribution

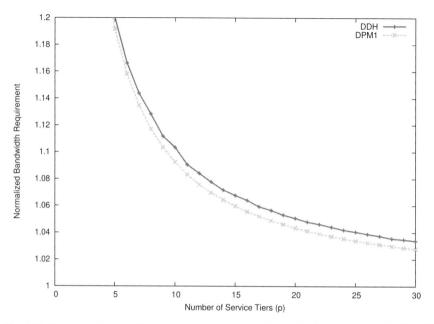

Fig. 4.10 Normalized bandwidth requirement against p, uniform distribution, $\beta = 0.05$ (© 2007 IEEE)

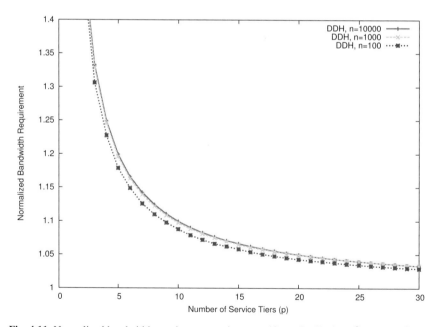

Fig. 4.11 Normalized bandwidth requirement against p, uniform distribution, $\beta = 0.05$ (© 2007 IEEE)

observed in the results we presented in the previous chapter, and can be explained by noting that as the number of levels becomes very large, the tiered-service network reduces to a continuous-rate network.

Fig. 4.11 shows the effect of the number n of demands on the normalized bandwidth requirement for the DDH algorithm. We observe that as the number n of demands increases, the normalized bandwidth requirement does increase slightly, but the effect diminishes quickly; in fact, the curve for $n = 10,000$ almost coincides with the curve for $n = 1,000$ in the figure. Overall, this behavior is similar to the one we observed for the DPM1 algorithm in Figs. 3.3-3.8.

Fig. 4.12 shows the impact of the overhead parameter β on the normalized bandwidth requirement for the DDH algorithm. Each point in the figure is the average over 30 problem instances with $n = 1000$ demand points generated from a uniform distribution. Five curves are shown, corresponding to $\beta = 0.001, 0.005, 0.01, 0.05$, and 0.9. It is clear from the figure that the value of the overhead parameter has little effect on the bandwidth penalty, and that all curves follow the familiar decreasing curve as the number p of supply points (i.e., service levels) increases.

The final figure of this section, Fig. 4.13, shows how the distribution of demand points affects the normalized bandwidth requirement metric for the DDH algorithm. Four curves are shown, corresponding to the triangle, increasing, decreasing, and uniform distributions, respectively. As we can see, the differences among the curves are pronounced when there is only a small number of supply points. This result is expected since, when p is small, the algorithm does not have sufficient flexibility to cover the demand points, hence the bandwidth penalty depends strongly on the given distribution. To illustrate this point, consider the special case when there is a single supply point, z_1. In order to satisfy the directionality constraints, it must then be that $z_1 = x_n$, where x_n is the largest demand point, and all other demand points are mapped to z_1. Since demand points are distributed in the interval $[0,1]$, for large n we can expect that $z_1 = x_n \approx 1$. Consider now the uniform distribution. The mean of n points generated from this distribution is approximately 0.5, but all n points are mapped to the single supply point whose value is approximately 1. Consequently, a tiered-service network is expected to require approximately twice the bandwidth of a continuous-rate one; Fig. 4.13 confirms this observation as the value of the curve for the uniform distribution for $p = 1$ is equal to 2. Similarly for the other distributions. On the other hand, as the number p of supply points increases, the DDH algorithm is able to position them near-optimally for each distribution, hence the differences among the various distributions start to disappear after about $p = 10$ supply points.

From the figures we presented in this section we can draw the following important conclusions regarding the performance of tiered service and TDM emulation from the network provider's point of view:

1. with $p = 10 - 15$ service levels, the bandwidth required by a tiered-service network is only about 5-10% higher than that of a continuous-rate network;
2. the additional constraint that all service levels be a multiple of a basic unit only slightly adds to the bandwidth penalty;

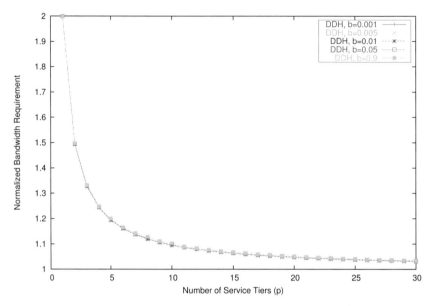

Fig. 4.12 Normalized bandwidth requirement against p for various values of β, uniform distribution, $n = 1000$

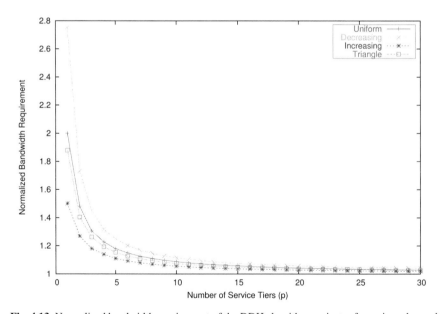

Fig. 4.13 Normalized bandwidth requirement of the DDH algorithm against p for various demand distributions, $n = 100$, $\beta = 0.05$

3. increasing the number n of demands (users) imposes only an incremental penalty on bandwidth; and, importantly,
4. the above observations are valid across a wide range of demand distributions.

4.2.3 Impact on Users: Blocking Probability

Let us now examine the practical impact of tiered service on the overall network performance as perceived by users. To this end, we consider an MPLS network scenario in which user requests in the form of label switched paths (LSPs) arrive and depart dynamically. An LSP between source-destination pair (s, d) requires a certain amount of bandwidth; if a path between s and d with sufficient resources can be found, the LSP is established, otherwise, it is rejected (blocked). The performance measure of interest in this context is the LSP blocking probability. We used simulation to compare the blocking probability of a continuous network to that of a tiered-service network. In a continuous network, an LSP requiring bandwidth x_i is accepted if a path with at least that much bandwidth can be found. In a tiered-service network, the bandwidth demand x_i is first mapped to the next highest service level offered, say, z_j, and the LSP is accepted if a path with bandwidth at least equal to z_j is found. The service levels for the tiered-service network are computed in advance for the given demand distribution, using the appropriate algorithm (DPM1 or a TDM-DPM1 heuristic).

In our simulation model, LSP requests arrive as a Poisson process with rate λ, and the mean LSP holding time is an exponentially distributed random variable with rate $\mu = 1$. Each simulation run lasts until 100,000 LSP requests have been served. Each point in the blocking probability curves shown here is the average of thirty simulation runs; we also plot 95% confidence intervals which we estimated using the method of batch means [101]. All the simulation results are for the 16-node network topology shown in Fig. 4.14 that was derived from the well-known 14-node NSFnet topology. All links of the network have unit cost, and LSPs are routed over the shortest (i.e., minimum-hop) path from the source to the destination. The capacity of all links is set to two units of bandwidth. Since the demand distributions in Table 3.3 are defined in the interval $[0, 1]$, this assumption implies that the bandwidth requested by any LSP is at most one half the link capacity.

Fig. 4.15 plots the blocking probability against the LSP arrival rate for a continuous network and two tiered-service networks, one using DPM1 to obtain the service levels and one using DDH, a representative algorithm for the TDM-DPM1 problem. As expected, the blocking probability of the continuous-rate network is lowest, that of the tiered-service network allocating bandwidth in multiples of a basic unit is highest (DDH algorithm), and that of a network which minimizes the excess bandwidth (DPM1 algorithm) is in between the other two. The higher blocking probability experienced by a tiered-service network is a direct result of the additional resources that such a network uses for each traffic demand. However, the increase

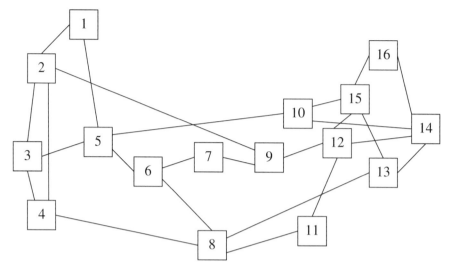

Fig. 4.14 The 16-node network topology used in the simulation experiments

in blocking probability is rather small and it may be more than compensated by the advantages of tiered service.

Fig. 4.16 shows the behavior of the blocking probability for the DDH algorithm as we vary the number of service levels p. The curves confirm the intuition that as p increases, the blocking probability of the tiered-service network decreases and tends towards that of a continuous-rate network. This figure suggests that the network designer/engineer may select the number p of the service levels to be offered so as to combine the advantages of tiered service with the performance of a continuous-rate one.

Fig. 4.17 illustrates the impact of the demand distribution on the blocking probability experienced by users. When traffic demands (i.e., LSP sizes) are drawn from the decreasing distribution, the blocking probability is the lowest. This fact can be explained by referring to the plot of the decreasing PDF in Fig. 3.2 and noting that the vast majority of the demands are small, with only a few of the demands approaching the maximum value of one-half the link capacity. Consequently, these small demands are relatively easy to accommodate (groom onto the network links), and the blocking probability is low. The reverse is true for the increasing distribution, under which most LSP requests are for bandwidth close to the maximum value. As a result, the blocking probability for the increasing distribution is quite high, up to two orders of magnitude higher than that of the decreasing distribution. The blocking probability curves for the other two distributions, triangle and uniform, lie between these two extremes. We conclude that the demand distribution has a profound effect on blocking probability. We emphasize, however, that this effect is due to the properties of the demand distributions, *not* to the tiered service *per se*. In fact, the blocking probability curves for a continuous-rate network (not shown

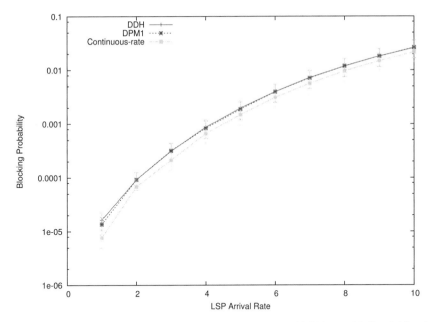

Fig. 4.15 Blocking probability against the LSP arrival rate, $n = 100,000$, $p = 30$, $\beta = 0.05$, uniform distribution (© 2007 IEEE)

here) are close to the corresponding curves of the tiered-service network (e.g., as illustrated in Fig. 4.15), and exhibit the same relative behavior as in Fig.4.17.

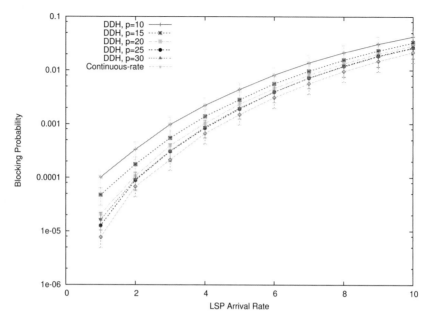

Fig. 4.16 Blocking probability against the LSP arrival rate, $n = 100,000$, $\beta = 0.05$, uniform distribution (© 2007 IEEE)

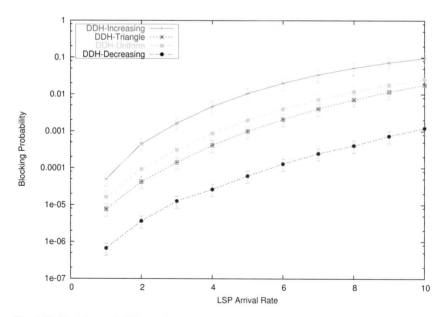

Fig. 4.17 Blocking probability against the LSP arrival rate, $n = 100,000$, $p = 30$, $\beta = 0.05$

Chapter 5
Bandwidth Tiered Service: Stochastic Demands

In Chapters 3 and 4 we considered the general problem of optimizing the bandwidth tier structure for tiered-service networks. The objective was to minimize the bandwidth penalty incurred by the provider relative to a continuous-rate network. An underlying assumption in these studies was that the size of the user population is fixed and does not change, and, furthermore, that the user demands are specified and known in advance. Consequently, the corresponding DPM1 problem and its variants include the fixed set X of demand points and their number n as part of the formulation.

In practice, within the problem domain we are considering, the user population is likely to undergo continuous changes, e.g., as users move in or out of the service area where a given provider operates. Customers are also likely to adjust their service over time. For instance, in a capacity-based tiering structure whereby the level of service is determined by Internet access speed, users may decide to upgrade to the next higher tier in response to the emergence of more bandwidth-intensive applications. Alternatively, with a usage-sensitive tiering structure, the service tier is a function of the amount of traffic generated by the user, hence it may vary from one billing period to another. In view of these observations, it is unlikely that providers will dimension their networks for a fixed set of bandwidth requests. Typically, network operators employ forecasting models to estimate the traffic demands over the medium- to long-term, and provision their services accordingly to match this traffic profile [65, 78].

In this chapter, we address the problem of determining the optimal set of service levels when only a probabilistic distribution of the size of users' bandwidth demands is known. Such a distribution may have been obtained empirically from existing user profiles, from predictive models, or a combination thereof. We show that this problem can be formulated as a variant of the directional p-median problem, and we present efficient algorithms to solve it.

G.N. Rouskas, *Internet Tiered Services*, DOI: 10.1007/978-0-387-09738-1_5,
© Springer Science + Business Media, LLC 2009

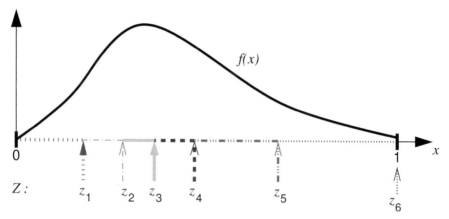

Fig. 5.1 Sample mapping from the domain of $f(x)$ to a solution set of 6 supply points (© 2003 IEEE)

5.1 The Stochastic Directional p-Median Problem

We now consider a variation of the directional p-median problem illustrated in Fig. 5.1. Let $f(x)$ and $F(x)$ be the probability density function (PDF) and cumulative distribution function (CDF), respectively, representing the population of demand points. Let x_{max} represent the maximum possible demand point. Without loss of generality, we make the assumption that all demand and supply points are normalized with respect to x_{max}. As a result, the domain of $f(x)$ and $F(x)$ is wholly contained within the interval $[0, 1]$.

Let μ be the mean of $f(x)$ and let $b \leq 1$ be the least upper bound on the domain of $f(x)$. A set $Z = \{z_1, \ldots, z_p\}$, $0 < z_1 < z_2 < \ldots < z_p < 1$, is a *feasible solution* of $f(x)$ if and only if $b \leq z_p$. For notational convenience, we let z_0 represent the "null" service tier, and set $z_0 = 0$. Note that each z_j is unique. Associated with a feasible solution Z is an implied *directional mapping* from the domain of $f(x)$ into Z, where $x \to z_j$ if and only if $z_{j-1} < x \leq z_j$. We may also write the implied mapping as

$$(x_{lower}, x_{upper}] \to z_j, \tag{5.1}$$

where $x_{lower} = z_{j-1}$ and $x_{upper} = z_j$.

The problem of determining the optimal service tiers when only the distribution of service demands is known, can be expressed as a variant of the directional p-median problem on the real line with deterministic demands that we studied in Chapter 3. We will refer to this new problem as SDPM1 (for "directional p-median problem in one dimension with stochastic demands"), and define it as follows.

Problem 5.1 (SDPM1). Given the PDF $f(x)$ and CDF $F(x)$ describing the population of demand points on the real line and an integer p, find a feasible set Z of p supply points, $Z = \{z_1, \ldots, z_p\}$, such that the following objective function is minimized:

$$T(z_1, \ldots, z_p) = \sum_{j=1}^{p} \left(\int_{z_{j-1}}^{z_j} (z_j - x) f(x) \, dx \right)$$

$$= \sum_{j=1}^{p} \left(z_j \int_{z_{j-1}}^{z_j} f(x) \, dx \right) - \mu \qquad (5.2)$$

where μ is the mean of $f(x)$.

Note that $T(z_1, \ldots, z_p)$ represents the *average* penalty, per demand, of excess bandwidth resources used by the tiered-service network than was requested by the original demand set. The term μ in (5.2) is the *average* amount of bandwidth requested by a demand point, while the summation term in the second line of (5.2) represents the *average* amount of bandwidth allocated to each demand point after the latter have been mapped to the appropriate supply points. In contrast, in the deterministic input case, the objective function $S(z_1, \ldots, z_p)$ of DPM1 in (2.16) is the *total* penalty under a tiered service for a particular demand set X, with the term $\sum_{j=1}^{p} (n_j z_j)$ representing the total bandwidth allocated to the demands and the term $[\rho_X = \sum_{i=1}^{n} x_i]$ corresponding to the total bandwidth of the original demand set.

We may derive the mathematical relationship between the objective functions of the DPM1 and SDPM1 problems as follows:

$$\lim_{n \to \infty} \frac{S(z_1, \ldots, z_p)}{n} = \lim_{n \to \infty} \left(\frac{1}{n} \left(\sum_{j=1}^{p} (n_j z_j) - \rho_X \right) \right)$$

$$= \lim_{n \to \infty} \left(\sum_{j=1}^{p} \left(\frac{n_j}{n} z_j \right) \right) - \lim_{n \to \infty} \left(\frac{\rho_X}{n} \right)$$

$$= \sum_{j=1}^{p} \left(z_j \lim_{n \to \infty} \left(\frac{n_j}{n} \right) \right) - \mu. \qquad (5.3)$$

Note that the limit of n_j/n as n goes to infinity equals the proportion of demand points x_i that fall within the interval (z_{j-1}, z_j), or $\int_{z_{j-1}}^{z_j} f(x) \, dx$. Thus we have:

$$\lim_{n \to \infty} \frac{S(z_1, \ldots, z_p)}{n} = \sum_{j=1}^{p} \left(z_j \int_{z_{j-1}}^{z_j} f(x) \, dx \right) - \mu$$

$$= T(z_1, \ldots, z_p). \qquad (5.4)$$

This last expression demonstrates that the DPM1 problem reduces to SDPM1 as the size of the demand set grows large.

Similarly, we may find an expression for the normalized bandwidth requirement metric for stochastic input, NBR_s, by taking the limit of expression (3.20) as n goes to infinity:

$$NBR_s = \lim_{n\to\infty} \frac{\sum_{j=1}^p (n_j z_j)}{\rho_X} = \lim_{n\to\infty} \frac{\sum_{j=1}^p \left(\frac{n_j}{n} z_j\right)}{\frac{\rho_X}{n}}$$

$$= \frac{\sum_{j=1}^p \left(z_j \lim_{n\to\infty}\left(\frac{n_j}{n}\right)\right)}{\lim_{n\to\infty}\left(\frac{\rho_X}{n}\right)} = \frac{\sum_{j=1}^p \left(z_j \int_{z_{j-1}}^{z_j} f(x)\,dx\right)}{\mu}. \qquad (5.5)$$

Because μ is a constant for a given $f(x)$, both the objective function $T(z_1,\ldots,z_p)$ and the normalized bandwidth requirement NBR_s are minimized whenever the numerator of the right-hand side of (5.5), i.e., the average bandwidth demand under tiered service, is minimized.

The following lemma is analogous to the fact, stated in Lemma 3.1, that, in the deterministic case, the largest supply point in an optimal supply set must equal the largest demand point x_n.

Lemma 5.1. *Let $f(x)$ be the PDF of the population of demand points on the real line, and $b \leq 1$ be the least upper bound on the domain of $f(x)$, as defined earlier. Let $Z = \{z_1,\ldots,z_p\}$, $0 < z_1 < z_2 < \ldots < z_p < 1$, be an optimal supply set for the corresponding SDPM1 problem instance. Then, $z_p = b$.*

Proof. By contradiction. Suppose $z_p \neq b$. From the definition of a feasible quantization set, we know that $b \leq z_p$; thus $b < z_p$. The values currently mapped to z_p lie in the interval $(z_{p-1}, b]$. Moving z_p down to b will reduce the objective function by a non-negligible amount equal to $\left[(z_p - b)\int_{z_{p-1}}^b f(x)dx\right]$. This contradicts the optimality of Z. Thus $z_p = b$. $\square$

5.2 Optimal Solution Through Nonlinear Programming

Rewriting the objective function $T(Z) = T(z_1,\ldots,z_p)$ from expression (5.2), we have the following optimization problem:

$$\text{Minimize}\quad T(Z) = \sum_{j=1}^p \left(z_j\left(F(z_j) - F(z_{j-1})\right)\right) - \mu \qquad (5.6)$$

$$\text{subject to}:\quad 0 < z_1 < z_2 < \ldots < z_{p-1} < z_p = b. \qquad (5.7)$$

From equation (5.6) it is clear that the objective function $T(Z)$ of the SDPM1 problem in not linear. We now describe a method that can be applied to instances of the SDPM1 problem for which the CDF $F(x)$ obeys the following two conditions:

1. $F(x)$ is twice differentiable, and
2. $F(x)$ is not piecewise defined over its entire domain.

In the following section, we present an approximate solution for instances of SDPM1 for which $F(x)$ fails to have these two properties.

When $F(x)$ is twice differentiable and not piecewise defined, $f(x)$ and its derivative $f'(x)$ are also not piecewise defined. Specifically, for each of $F(x)$, $f(x)$, and $f'(x)$, it is possible to write the function as a single closed form expression over its entire domain, a necessary property for applying the following method [13]: locate a critical point of $T(Z)$ and then verify that the point is a minimum.

To find a critical point, we set the first order partial derivatives of $T(Z)$ with respect to $z_j, j = 1,\ldots,p-1$, equal to zero, yielding a set of $p-1$ simultaneous differential equations in $p-1$ unknowns. The highest supply point z_p is known: from Lemma 5.1 we know that $z_p = b$. It will then be possible to solve for each $z_j, j = 2,\ldots,p$, in terms of z_1 only. Since $z_p = b$, we can find z_1. Through back-substitution, we can then obtain the remaining values for $z_j, j = 2,\ldots,p-1$.

Taking the partial derivative of $T(Z)$ with respect to $z_j, j = 1,\ldots,p-1$, we have:

$$\frac{\partial T(Z)}{\partial z_j} = z_j \frac{\partial F(z_j)}{\partial z_j} + \left(F(z_j) - F(z_{j-1})\right) - z_{j+1} \frac{\partial F(z_j)}{\partial z_j}$$

$$= (z_j - z_{j+1}) \frac{\partial F(z_j)}{\partial z_j} + F(z_j) - F(z_{j-1})$$

$$= (z_j - z_{j+1})f(z_j) + F(z_j) - F(z_{j-1}) \qquad (5.8)$$

From the equation $\frac{\partial T(Z)}{\partial z_j} = 0, j = 1,\ldots,p-1$, we can solve for z_{j+1} in terms of z_j and z_{j-1}, yielding the following recursive expression:

$$z_{j+1} = z_j + \frac{F(z_j) - F(z_{j-1})}{f(z_j)}, \quad j = 1,\ldots,p-1 \qquad (5.9)$$

Since $z_0 = 0$[1], then $F(z_0) = 0$. For the equation corresponding to $\frac{\partial T(Z)}{\partial z_1} = 0$, we have:

$$z_2 = z_1 + \frac{F(z_1)}{f(z_1)} \qquad (5.10)$$

Thus we have obtained z_2 in terms of z_1 only. For the equation corresponding to $\frac{\partial T(Z)}{\partial z_2} = 0$, we have:

$$z_3 = z_2 + \frac{F(z_2) - F(z_1)}{f(z_2)} \qquad (5.11)$$

---

[1] Recall that z_0 was defined as the "null" service tier.

Using Equation (5.10) to substitute for z_2 in Equation (5.11) above, gives an expression for z_3 in terms of z_1 only. For the equation corresponding to $\frac{\partial T(Z)}{\partial z_3} = 0$, we have:

$$z_4 = z_3 + \frac{F(z_3) - F(z_2)}{f(z_3)} \tag{5.12}$$

Since we already have both z_3 and z_2 in terms of z_1 only, we can use substitution to get z_4 in terms of z_1 only. In general, we can obtain an expression for z_{j+1} in terms of z_1 only, after using substitution in the equation corresponding to $\frac{\partial T(Z)}{\partial z_j} = 0$.

The final equation, corresponding to $\frac{\partial T(Z)}{\partial z_{p-1}} = 0$, is:

$$b = z_p = z_{p-1} + \frac{F(z_{p-1}) - F(z_{p-2})}{f(z_{p-1})} \tag{5.13}$$

After substitution, the left-hand side of this equation is the constant b, and the right-hand side is a function of z_1. Thus we can solve for z_1. All other values of $z_j, j = 2, \ldots, p-1$, can be obtained once z_1 is known.

Notice that the feasible region, defined by $0 < z_1 < z_2 < \ldots < z_{p-1} < z_p = b$, is a convex set. If $F(x)$ is a convex function, then $T(Z)$ is also convex, and the critical point $(z_1, z_2, \ldots, z_{p-1})$ obtained from the above method will be a global minimum. Otherwise, the critical point $(z_1, z_2, \ldots, z_{p-1})$ is a global minimum if and only if the Hessian matrix of second partial derivatives of $T(Z)$ is positive definite. Since the Hessian for $T(Z)$ turns out to be a symmetric tridiagonal matrix, it can be shown to be positive definite (or not) in time $O(p^2)$ [30].

5.2.1 Example: Solution for the Uniform Demand Distribution

Due to the simplicity of the uniform distribution, namely $f(x) = 1$ and $F(x) = x$, it is possible to solve for the optimal values of $z_1, \ldots, z_{p-1}$, without specifying a particular value for p. The domain of the uniform distribution is $[0, 1]$, thus from Lemma 5.1 we have $z_p = 1$. Using equation (5.9) we obtain:

$$z_{j+1} = z_j + \frac{z_j - z_{j-1}}{1} = 2z_j - z_{j-1}, \quad j = 1, \ldots, p-1. \tag{5.14}$$

Recalling that $z_0 = 0$, the first equation (corresponding to $\frac{\partial T}{\partial z_1} = 0$) yields: $z_2 = 2z_1$. From the second equation we have: $z_3 = 2z_2 - z_1 = 3z_1$; from the third: $z_4 = 2z_3 - z_2 = 4z_1$; and so on, up to the $(p-1)$-th equation: $z_p = pz_1$. In general, $z_j = jz_1$ for $z = 2, \ldots, p$. Using the additional information that $z_p = 1$, we have that $z_p = pz_1 = 1$. Thus, $z_1 = \frac{1}{p}$, yielding the solution:

$$z_j = \frac{j}{p}, \quad j = 1, \ldots, p, \tag{5.15}$$

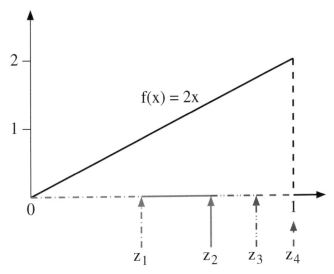

Fig. 5.2 Optimal solution to the SDPM1 instance for the increasing distribution and $p = 4$ supply points.

which implies that the supply points are uniformly located along the interval $[0,1]$, as expected.

5.2.2 Example: Solution for the Increasing Demand Distribution

As another example, consider the increasing distribution with $f(x) = 2x$ and $F(x) = x^2$, shown in Table 3.3 and Fig. 3.2. $F(x)$ is a convex function, therefore the critical point $(z_1, \ldots, z_p)$ obtained from the method we described above is the global minimum. From the recursive expression (5.9) we obtain:

$$z_{j+1} = z_j + \frac{z_j^2 - z_{j-1}^2}{2z_j} = \frac{3z_j^2 - z_{j-1}^2}{2z_j}, \quad j = 1, \ldots, p-1. \qquad (5.16)$$

For $j = 1$, and since $z_0 = 0$, equation (5.16) yields: $z_2 = \frac{3}{2}z_1$. By substituting this value of z_2 into the equation for $j = 2$, we obtain z_3 as a function of z_1 only: $z_3 = \frac{23}{12}z_1$. Continuing in this manner, the equation for $j = 3$ gives $z_4 = \frac{111}{46}z_1$. Now, assuming that there are $p = 4$ supply points, and using the fact that $z_4 = 1$, we can use these last three equations to obtain the critical point:

$$z_1 = \frac{46}{111}, \quad z_2 = \frac{69}{111}, \quad z_3 = \frac{529}{666}, \quad z_4 = 1. \qquad (5.17)$$

Fig. 5.2 plots the PDF of the increasing distribution and the optimal solution to the corresponding SDPM1 problem instance with $p = 4$ supply points. As we can see, the optimal supply points are located towards the rightmost part of the interval $[0, 1]$, reflecting the fact that the mass of the distribution is concentrated in that area. Comparing to the uniform distribution, we conclude that the optimal solution to SDPM1 is strongly dependent on the input demand distribution.

5.3 An Efficient Approximate Solution

In general, a given demand distribution function $F(x)$ may not satisfy the two conditions listed in the previous section, or it may not be convex, in which case the objective function (5.6) will not be convex either. For instance, an empirically obtained CDF F(x) may not be continuous, in which case the Hessian matrix is not defined. In such a case, it may be possible to formulate and solve approximate linear programming formulations of the non-linear optimization problem defined by equations (5.6) and (5.7), or apply branch-and-bound techniques [13]. One drawback of such solution methods is that they have to be customized to the input distribution function. More importantly, such methods may need a large number of iterations, or may get trapped at a local maximum.

Another option would be to fit $F(x)$ (e.g., using moment-matching techniques [32, 69]) to a function $\hat{F}(x)$ that satisfies the two conditions, and then use the technique we presented in the previous section on $\hat{F}(x)$. There are two issues with such an approach. First, identifying an appropriate function $\hat{F}(x)$ is a non-trivial task, and may even be a difficult problem for an arbitrary distribution $F(x)$. Second, even if a function $\hat{F}(x)$ that satisfies the two conditions above and is a good fit (based on some appropriate metric) for $F(x)$ could be found, it would be difficult to quantify the error introduced in the optimal solution by using $\hat{F}(x)$ instead of $F(x)$.

In order to overcome these challenges, we present next an approximate yet accurate method for solving general instances of the SDPM1 problem, in which $F(x)$ may be any arbitrary cumulative distribution function that need not be convex or even continuous. Rather than developing a sub-optimal algorithm for solving SDPM1 directly, we take a different approach: we provide an approximate formulation of SDPM1 that asymptotically converges to the formulation in (5.6)-(5.7), along with an algorithm that solves this new problem optimally.

5.3.1 An Approximate Formulation of SDPM1

We note that it is always possible to create a discrete approximation of the PDF $f(x)$, regardless of its form, as illustrated in Fig. 5.3. In particular, we can choose an integer $K > p$, where p is the number of supply points, and partition the interval

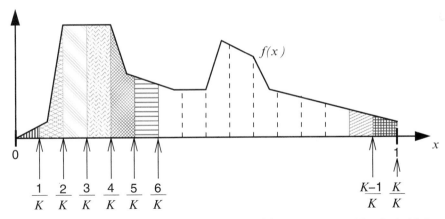

Fig. 5.3 Forming a PDF approximation: The area under $f(x)$ over a sub-interval is paired with the right-hand endpoint of the sub-interval (© 2003 IEEE)

$[0,1]$ into K sub-intervals each of length equal[2] to $\frac{1}{K}$. The right-hand endpoint of the k-th interval is $e_k = \frac{k}{K}$, and we let $e_0 = 0$. We associate with endpoint e_k a discrete point mass density

$$P_k = \int_{e_{k-1}}^{e_k} f(x)\, dx, \quad k = 1,\ldots,K. \tag{5.18}$$

The K pairs $\{(e_k, P_k)\}$ form the approximation of $f(x)$. We also define

$$F_k = \sum_{i=1}^{k} P_i, \quad k = 1,\cdots,K, \tag{5.19}$$

so that the K pairs $\{(e_k, F_k)\}$ form the approximation of the CDF $F(x)$.

Our approximate approach to solving SDPM1 is based on the observation that one can view the k-th ordered pair as representing P_k demand points each with the value e_k. More specifically, the demand points for this instance of the problem take values from the discrete set $\{e_k, k = 1,\ldots,K\}$ of sub-interval endpoints. Then, the penalty of mapping the P_k demand points e_k to another point $e_l > e_k$ is $P_k \times (e_l - e_k)$. Since the demand set is discrete, from Lemma 3.1 we have that the p optimal supply points also take values from the set $\{e_k\}$. Consequently, we can formulate the following approximate version of SDPM1 as a discrete problem that is similar to DPM1:

---

[2] The K sub-intervals do not need to be of equal length, although we make this assumption here for simplicity. In Section 5.3.3 we discuss scenarios in which it might be beneficial to employ sub-intervals of unequal length, e.g., so as to align them with the points of discontinuity of a given demand distribution.

Problem 5.2 (Approximate-SDPM1). Given the the K-point approximation $\{e_k, P_k\}$ of the PDF of demand points and an integer number $p < K$ of supply points, find a feasible set Z of supply points, $z_1 < z_2 < \cdots < z_p$, that minimizes the objective function:

$$\bar{T}(Z) = \bar{T}(z_1, \cdots, z_p) = \sum_{j=1}^{p} \left(z_j \left(F_{k_j} - F_{k_{j-1}} \right) \right) \qquad (5.20)$$

subject to the constraints:

$$z_j = e_{k_j} \in \{e_k\}, \quad j = 1, \cdots, p, \; k = 1, \cdots, K \qquad (5.21)$$

$$z_1 < z_2 < \cdots < z_p. \qquad (5.22)$$

The objective function $\bar{T}(Z)$ in (5.20) represents the total *approximate expected bandwidth* required to satisfy the demands under the tiered service. Indeed, the term $\left(F_{k_j} - F_{k_{j-1}} \right)$ is the fraction of demand points falling in the interval $(z_{j-1}, z_j]$ and are mapped to z_j. Hence, the j-th term of $\bar{T}(Z)$ is the total expected bandwidth required for the demands mapped to supply point z_j. We also note that, while the objective function $T(Z)$ of SDPM1 in 5.2 includes a term corresponding to the mean μ of the PDF $f(x)$, the function $\bar{T}(Z)$ does not contain such a term. Since μ is a constant for a given instance of this problem, including it does not affect the optimal solution, hence we have omitted it from $\bar{T}(Z)$.

Clearly, as $K \to \infty$, the PDF approximation approaches the original PDF and Approximate-SDPM1 reduces to the original SDPM1 problem.

5.3.2 Optimal Solution to Approximate-SDPM1

In order to solve Approximate-SDPM1, we define $\Phi(k, l)$ as the optimal value of the objective function (5.20) when the number of intervals in the PDF approximation is k and the number of supply points is $l \leq k$. Then, $\Phi(k, l)$ may be computed recursively as follows:

$$\Phi(k, 1) = e_k F_k, \quad k = 1, \cdots, K \qquad (5.23)$$
$$\Phi(k, l+1) = \max_{q=l,\cdots,k-1} \{\Phi(q, l) + e_k(F_k - F_q)\},$$
$$l = 1, \cdots, p-1; \; k = 2, \cdots, K \qquad (5.24)$$

The boundary conditions (5.23) can be explained by observing that if there is only one tier of service, it must coincide with the right-hand endpoint of the k-th (i.e., rightmost) interval. The recursive expression (5.24) simply states that, for $l+1$ service tiers, the largest tier must coincide with the right-hand endpoint of the k-th interval, and the remaining l tiers must be optimally assigned to the endpoints of any feasible interval $q, l \leq q \leq k-1$.

The running time of the above dynamic programming algorithm to obtain $\Phi(K,p)$ is $O(pK^2)$. Note that $\Phi(K,p)$ is the optimal solution to Approximate-SDPM1 with p supply points and K intervals, which in turn is an approximate formulation of the original SDPM1 problem. Hence, $\Phi(K,p)$ represents a solution close to the optimal solution to SDPM1, regardless of the shape of the objective function (5.2) or the form of the corresponding PDF $f(x)$. Clearly, the better the PDF approximation, i.e., the larger the value of K, the closer that $\Phi(K,p)$ will be to the true optimal solution for a given PDF; the tradeoff is an increase in running time. As we demonstrate in the subsection, the value of $\Phi(K,p)$ converges quickly as the value of K approaches 50-100 for all the distribution functions we have considered, thus a (near-) optimal solution can be computed efficiently for any instance of SDPM1.

An interesting observation is that the solution obtained by the dynamic programming algorithm (5.23)-(5.24) consists of supply points whose values $z_j, j = 1, \ldots, p$, are multiples of the same unit $1/K$. Hence, this solution falls within the class of solutions for the TDM emulation problem we investigated in Chapter 4, with $1/K$ representing the unit of bandwidth allocation.

5.3.3 *Convergence of the Approximate Solution*

As we mentioned earlier, the quality of the approximate solution $\Phi(K,p)$ to the SDPM1 problem improves as the number K of intervals in the approximation of the demand distribution $f(x)$ increases; the downside is a corresponding quadratic increase in the running time of the dynamic programming algorithm. In this section we investigate the rate of convergence of the approximate solution as a function of K. To this end, we ran the algorithm (5.23)-(5.24) on the six demand distributions described in Section 3.3 and listed in Table 3.3 for a range of values of K and p. The performance metric we used to evaluate the relative performance of the solutions for the various values of K is the normalized bandwidth requirement for stochastic input, NBR_s, shown in equation (5.5). This normalized measure makes it possible to readily compare results among these very different distributions.

The six figures 5.4-5.9 plot the value of the NBR_s metric against K for each of the six demand distributions in Table 3.3, respectively. For the experiments shown here, we let K to take on the values from 10 to 100 in increments of 5. Each figure contains 7 curves, each corresponding to a different value of p from the set $\{5, 10, 15, 20, 25, 30, 35\}$. Note that, as we discussed above, the p supply points are selected among the K endpoints of the PDF approximation; hence the dynamic programming algorithm (5.23)-(5.24) is defined only for $K > p$. As a result, curves for higher values of p in the six figures start at correspondingly larger values of K, so as to ensure that this requirement is met.

We make two important observations from these figures. First, within a given demand distribution and for a fixed value of the number K of intervals, as the number p of supply points increases, the value of the NBR_s decreases and approaches 1. Equivalently, the curve for a given value of p within a figure lies above the curve for a value $p' > p$. This is the same behavior we observed in earlier chapters, and reflects the fact that as p grows large, the tiered-service system approaches a continuous-rate one and the penalty due to tiered-service diminishes. Consistent with intuition, the figures also demonstrate the effect of diminishing returns, as further increases in the number p of supply points provide smaller improvements to the value of NBR_s.

The second observation has to do with the rate of convergence of the NBR_s measure as the quality of the PDF approximation increases. Specifically, for a particular value of p, as K increases, the value of NBR_s either decreases or oscillates slightly, depending on the demand distribution. Nevertheless, in either case the curves quickly settle down to a particular value that depends on p and the distribution. For example, in Fig. 5.5 (triangle distribution), the curve for $p = 20$ settles down to a value of $NBR_s \approx 1.045$ as early as $K = 25$. In other words, by dividing the interval $[0, 1]$ into as few as 25 sub-intervals of equal length, we can accurately estimate the effect on required resources due to a tiering structure consisting of 20 levels.

Although the convergence behavior we just described is generally consistent across the various distributions and values of p, some of the curves in Fig. 5.8 (unimodal distribution) and Fig. 5.9 (bimodal distribution) do not appear to settle down as quickly within this range of values for K. Instead the curves exhibit an intriguing sinusoidal shape. Interestingly, as shown in Fig. 3.2, the unimodal and bimodal distributions are the two distributions among the set that possess sharp discontinuities. To investigate further, we generated another set of graphs, shown in Fig 5.10, for the bimodal distribution, this time letting K take on the values $10, 15, 20, \ldots, 300$. We observe that from $K = 100$ to $K = 300$, the sinusoidal shape quickly decreases in amplitude and settles down to a particular value of NBR_s. Looking closer at the behavior of the curves for the various values of K, we have concluded that the sinusoidal shape can be attributed to the location of the endpoints of the K intervals, and in particular, whether these endpoints adequately fall along the points of discontinuity of the PDF. Recall from Table 3.3 that the first peak in the bimodal distribution rises at $x = .25$ and falls at $x = .35$, and the second peak rises at $x = .65$ and falls at $x = .75$. Hence, when $K = 20, 40, 60, \ldots$, there are endpoints e_k that exactly equal .25, .35, .65, and .75; as a result, these K values correspond to the valleys (lower values of NBR_s) along the various curves in Figs. 5.9 and 5.10. On the other hand, the peaks of the curves occur at values of K for which the interval endpoints do not align well with the points of discontinuity of the bimodal distribution.

The main conclusion to draw from these observations is that a probability density function $f(x)$ with discontinuities will be better approximated (and hence, the dynamic programming algorithm $\Phi(K, p)$ will perform better) whenever the K intervals are chosen such that their endpoints lie at the points of discontinuity. In fact, the operation of the recursion $\Phi(K, p)$ does not require that the input pairs (e_k, P_k) be evenly spaced along the interval $[0, 1]$. Therefore, whenever $f(x)$ has many dis-

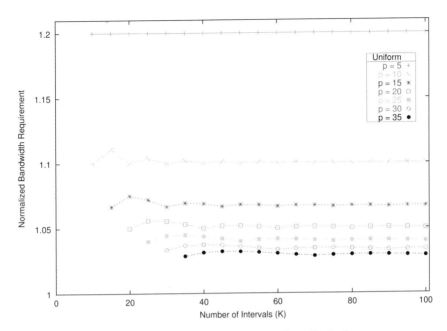

Fig. 5.4 Normalized bandwidth requirement NBR_s vs. K, uniform distribution

continuities, the endpoints e_k may be particularly chosen to fall at the points of discontinuity to achieve better performance from $\Phi(K,p)$.

Overall, our experiments indicate that $K \approx 100 - 200$ is sufficient for convergence, confirming that the dynamic programming algorithm may obtain accurate solutions to the original SDPM1 problem efficiently.

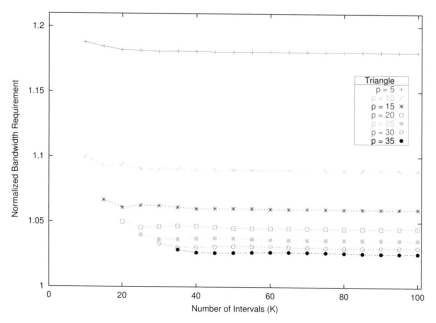

Fig. 5.5 Normalized bandwidth requirement NBR_s vs. K, triangle distribution (© 2003 IEEE)

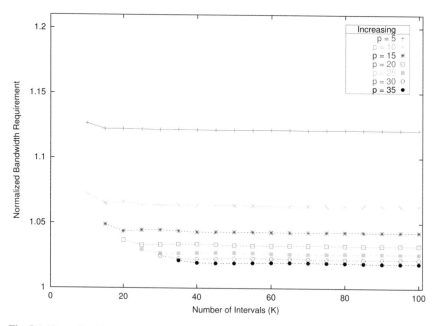

Fig. 5.6 Normalized bandwidth requirement NBR_s vs. K, increasing distribution

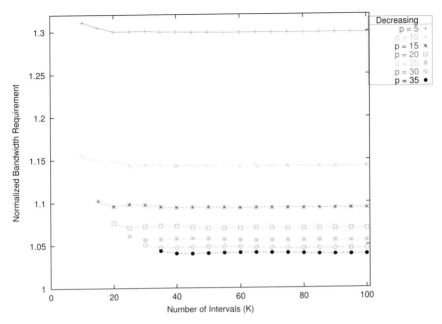

Fig. 5.7 Normalized bandwidth requirement NBR_s vs. K, decreasing distribution

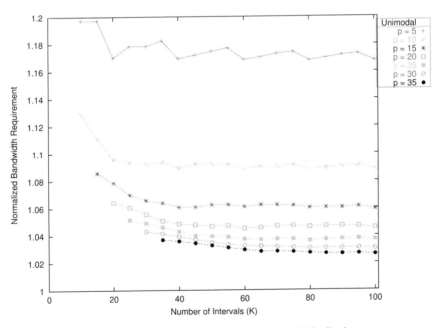

Fig. 5.8 Normalized bandwidth requirement NBR_s vs. K, unimodal distribution

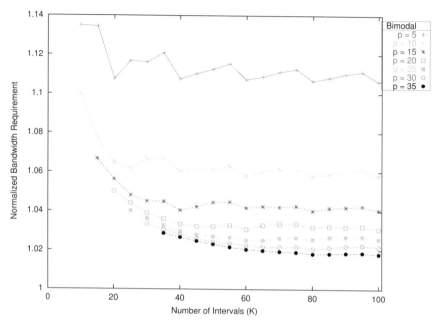

Fig. 5.9 Normalized bandwidth requirement NBR_s vs. K, bimodal distribution (ⓒ 2003 IEEE)

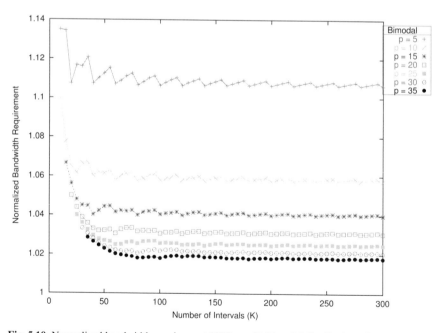

Fig. 5.10 Normalized bandwidth requirement NBR_s vs. K, bimodal distribution (ⓒ 2003 IEEE)

Chapter 6
Tiered Structures for Multiple Services

In our study of tiered service so far, we have considered only the case where the network service is characterized by a single parameter, e.g., Internet access speed or amount of user-generated traffic. In practice, network providers often package several distinct services into a *bundle*, or provide a single service that may be characterized by multiple parameters. Service bundles, and associated tiered structures, are quite common in the telecommunications market. For instance, wireless providers combine voice, data, and text services into tiered subscription packages marketed to users, where a given tier corresponds to a certain combination of values for voice minutes, Internet (e.g., web and email) data, and text messages available to the user during the billing period (with additional charges applying if the user exceeds its allocation in any of the services). Similarly, ISPs may bundle a broadband access service with an email or web hosting service (for which fees may be based on the amount of traffic handled), and possibly an online storage service (characterized by the amount of data the user may store on the provider's servers).

Alternatively, a single network service may be characterized by multiple parameters. Although typical residential Internet access services are marketed based solely on access speed, business clients may also be concerned with reliability (e.g., the level of protection of the access link), the transfer delay of their data, or the response time experience by their own customers (e.g., for web hosting services). The level of expectations in terms of the various quality of service (QoS) parameters are described in detail in the service level agreements (SLAs) negotiated between the network provider and business clients.

Based on these observations, it is desirable to design multi-dimensional tiered structures, where each dimension corresponds to a certain level of one distinct service (for service bundles) or one service parameter (for services characterized by multiple QoS parameters). The objective in this case would be to determine a tiered structure that is *jointly optimal* for a vector of services or service parameters. With such a tiered structure, a user with a certain level of requirements for each service parameter would subscribe to the tier that offers a level at least equal to its requirement across all dimensions of service. From the network provider's point of view, mapping a user to the next highest available tier involves the provision of additional

G.N. Rouskas, *Internet Tiered Services*, DOI: 10.1007/978-0-387-09738-1_6,
© Springer Science + Business Media, LLC 2009

resources relative to a network that exactly matches the users' requirements in terms of each service. Therefore, this problem of *vector quantization* can be modeled as a directional p-median problem in multiple dimensions. Note also that in this case, quantization (i.e., service tiering) is even more important since the space of potential service levels grows as the product of the space for each service parameter.

In this chapter, we introduce the directional p-median problem on the plane as the fundamental problem for studying multi-dimensional tiered-service structures. We show that the problem is NP-complete, and we develop a new heuristic to solve it. We also design an efficient dynamic programming algorithm that finds the optimal set of tiers from within a special class of tiered structures that are important in practice. Although our discussion in the rest of this chapter is limited to tiered structures with only two dimensions, the results and algorithms presented here may be extended in a straightforward manner to more than two dimensions.

6.1 The Directional p-Median Problem on the Plane

Formally, we define the directional rectilinear p-median problem in two dimensions (problem DPM2) to be:

Problem 6.1 (DPM2). Given a set $X = \{(x_1,y_1),(x_2,y_2),\ldots,(x_n,y_n)\}$ of n demand points in the plane and an integer p, find a feasible set $Z = \{(z_1,t_1),(z_2,t_2),\ldots,(z_p,t_p)\}$ of p supply points such that the following objective function is minimized:

$$Q(Z) = \sum_{i=1}^{n} \min_{1 \le j \le p} \left\{ D_{dr}^{(2,2)}((x_i,y_i),(z_j,t_j)) \right\} \qquad (6.1)$$

where the 2-directional rectilinear distance metric $D_{dr}^{(2,2)}$ is defined in expression (2.17).

We have the following result.

Theorem 6.1. *The decision version of problem DPM2 is NP-complete.*

Proof. The reduction is from planar 3-SAT, the non-polar version, first proved NP-complete by Lichtenstein [75]. The details of the proof are omitted; the interested reader is referred to [59, Chapter 4] for the full proof. Note also that this result implies NP-completeness of the directional p-median problem in more than two dimensions. □

Let $X_j, j = 1,\ldots,p$, denote the set of demand points mapped to the j-th supply point (z_j,t_j), and let n_j denote the cardinality of X_j. Sets X_j are pairwise disjoint and such that $\bigcup_{j=1}^{p} X_j = X$. Because of the directionality constraints, we have that:

$$\forall (x,y) \in X_j : \; z_j \geq x \text{ and } t_j \geq y, \quad j = 1, \ldots, p. \tag{6.2}$$

Therefore, we can rewrite the DPM2 objective function (6.1) as:

$$
\begin{aligned}
Q(Z) &= \sum_{j=1}^{p} \sum_{(x,y) \in X_j} [(z_j - x) + (t_j - y)] \\
&= \sum_{j=1}^{p} [n_j(z_j + t_j)] - \sum_{i=1}^{n} (x_i + y_i) \\
&= \sum_{j=1}^{p} [n_j(z_j + t_j)] - \rho_X
\end{aligned}
\tag{6.3}
$$

where ρ_X represents the total demand in the input set X, and is a generalization of the similar quantity we defined in Chapter 3.3 for the single-dimensional case.

In this chapter, we only consider the discrete version of the DPM2 problem as defined above, we develop efficient heuristic algorithms, and we evaluate the quality of the resulting solutions through simulations. This version of DPM2 arises in applications where the user population is fixed and demands are deterministic and known in advance. The problem definition can be extended to model stochastic demands (e.g., similar to SDPM1 in Chapter 5), or to impose additional constraints on the solution set (e.g., similar to TDM emulation in Chapter 4). The techniques we developed in those chapters for the single-dimensional problem variants can be easily adapted to tackle the corresponding variants in multiple dimensions.

6.2 Heuristic Algorithms for Discrete-PM2

Before proceeding to tackle DPM2, let us take a closer look at the discrete p-median problem on the plane, Discrete-PM2, whose formulation is provided in Chapter 2. As we mentioned there, Discrete-PM2 is NP-hard. Nevertheless, this problem arises naturally in a wide array of facility location problems, and hence it is of high practical importance to many application domains. Consequently, it has been extensively studied over several decades, both in its classical form and in numerous variants. Here, we briefly review results that pertain to our later study of DPM2. We first discuss the impact of the distance measure on the difficulty of a given problem instance, as measured by the amount of computational effort required to obtain good solutions, as well as the implications for the directional distance metric that is of interest to us. We then describe several heuristics for Discrete-PM2 and discuss their applicability to the DPM2 problem.

6.2.1 Effect of Distance Properties on Computational Effort

Recall from Chapter 2.1.1 that the supply points comprising a solution to Discrete-PM2 can only be selected among a set of *candidate* points. Let n be the number of demand points and m the number of candidate points. The input to Discrete-PM2 includes an $n \times m$ distance matrix $[d_{ij}]$ that holds the distances from demand points to candidate points. Different distance metrics produce distance matrices with different properties that influence the performance of a heuristic. Two properties whose impact on performance has been investigated are (1) symmetry and (2) the ability to satisfy the triangle inequality. According to the study by Schilling *et al.* [104], the lack of symmetry has a minor impact on performance, but failure to satisfy the triangle inequality is a much graver crime, making optimal or even good quality solutions hard to find. Note that (non-directional) rectilinear and Euclidean distance matrices both are symmetric and obey the triangle inequality. On the other hand, a randomly generated distance matrix in general will neither be symmetric nor obey the triangle inequality. As for the directional rectilinear distance metric, the outlook is positive as regards the more critical attribute, since the distance matrix obeys the triangle inequality. However, the matrix is not symmetric: from the definition (2.17) of 2-directional rectilinear distance, if $D_{dr}^{(2,2)}((x_i, y_i), (x_j, y_j))$ is finite and > 0, then $D_{dr}^{(2,2)}((x_j, y_j), (x_i, y_i))$ is infinite.

ReVelle [95] labels certain Discrete-PM2 problems as *integer friendly*, meaning that "either integer termination of linear programming formulations are frequent, or little branch and bound is needed to resolve the problem in integers." The main conclusion of [104] is that the property of obeying the triangle inequality has the greatest impact on the integer friendliness of an instance of Discrete-PM2. The question of why this is the case remains open.

6.2.2 Teitz and Bart (TB) Vertex Substitution Heuristic

The 1968 Teitz and Bart (TB) [111] vertex substitution heuristic for the Discrete-PM2 problem is well-known and much studied. A study comparing TB to exact methods has shown that TB rarely becomes trapped in local minima [99]. TB can be easily adapted for the directional problem.

A pseudocode description of the TB heuristic is provided in Fig. 6.1. The algorithm begins with an initial solution of p supply points which are numbered arbitrarily from 1 to p. Assigning each demand point to its nearest supply point, the heuristic evaluates the objective function for this solution. Next, the algorithm enters the first major iteration (Steps 3-8 in Fig. 6.1). Within this iteration, the first supply point is replaced with the candidate point (not already in the solution) that causes the greatest decrease in the objective function. This process is repeated by examining each of the remaining supply points in turn. Each time, the heuristic seeks the best candidate to replace the supply point being considered for removal, given that all other supply points in the solution are fixed. The first major iteration ends when the heuristic has tried removing each of the p solution points, and the final solution becomes the initial solution for the second major iteration. TB terminates when a major iteration results in no changes to the solution, usually within only a few (≤ 5) iterations. Fig. 6.2 illustrates the operation of the TB heuristic during one major iteration on a sample problem instance; note that in this example we use the directional rectilinear distance metric to demonstrate that TB can be readily used for solving the DPM2 problem.

In each major iteration of the TB heuristic, the **for** loop between Steps 4-7 in Fig. 6.1 is executed p times, once for each supply point. Selecting a candidate point to replace a given supply point in Step 5 of the algorithm takes time $O(n(m-p))$, where m is the number of candidate points: there are $m - p$ alternate candidate points, and for each we need to examine all of the n demand points to determine the amount of change in the objective function. Hence, each major iteration of TB runs in time $O(np(m-p))$. Recall from Chapter 2 that the number m of candidate points is $O(n^2)$ for both the Discrete-PM2 and the DPM2 problems. Therefore, the worst-case running time of each major iteration of TB is $O(n^3p)$.

Clearly, the performance of TB depends on the starting solution. Hence, a common approach to obtaining good solutions is to generate a number of random starting solutions as input for multiple TB runs, and then to choose the best solution from among the local optima that are found.

6.2.3 The Global/Regional Interchange Algorithm (GRIA)

The Densham and Rushton global/regional interchange algorithm (GRIA) improves the runtime of TB by taking advantage of characteristics of five well-known heuristics for Discrete-PM2 [29]. Speedup is achieved through the use of novel data structures that enable the heuristic to evaluate far fewer supply point substitutions. GRIA iterates between a global and a regional phase until an iteration of both phases yields no change. Applied to either Discrete-PM2 or DPM2, the global phase terminates with a feasible solution, i.e., one in which each demand point is assigned to its "nearest" supply point (according to the appropriate distance metric). Next, during the regional phase, each subset of points assigned to a supply point is considered as a separate 1-median problem and is solved independently. For Discrete-PM2, the solution z' to this 1-median problem could be different from the original supply point z, in which case making an exchange (i.e., replacing z with z') would improve the overall objective function.

However, if GRIA is applied to an instance of DPM2, no exchanges will ever take place during the regional phase, since there will never be more than one valid directional median for a given subset of demand points. The only feasible 1-median for a subset of points is the point (x_{max}, y_{max}), where x_{max} (respectively, y_{max}), is the maximum-valued x (respectively, y), from among all points in the subset. As a

Teitz and Bart Vertex Substitution Heuristic (TB)
Input: A demand set X of size n, the set of candidate points C of size m, the number of supply points p, and the $n \times m$ distance matrix $[d_{ij}]$.
Output: A set Z of supply points.

begin
1. $Z \leftarrow$ initial set $\{(z_1,t_1),\ldots,(z_p,t_p)\}$ of supply points randomly selected from the set C
2. $S \leftarrow$ objective function value for set Z
3.   **repeat** // major iteration
4.   **for** $i \leftarrow 1$ **to** p **do**
5. $(z_i,t_i) \leftarrow$ candidate point in $C \setminus Z$ that results in greatest decrease in objective function S
6.   Update the objective function S
7.   **endfor**
8.   **until** no change in set Z
9.   **return** Z
end

Fig. 6.1 The TB heuristic.

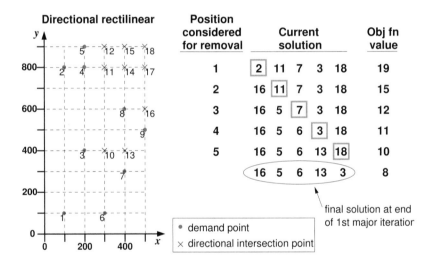

Fig. 6.2 One major iteration of the TB heuristic on a DPM2 problem instance with $n = 9$ demand points, $p = 5$ supply points, and $m = 18$ candidate points.

result, the regional phase will always fail to cause an exchange, hence GRIA cannot be used to solve DPM2.

6.2.4 Heuristic Concentration (HC)

Heuristic concentration (HC) is a two-stage algorithm developed by Rosing and ReVelle [100]. Although they demonstrate how HC works by applying it to Discrete-PM2, HC is a metaheuristic [88] that can be applied to many different combinatorial problems. HC attempts to glean information from the many local minima obtained from repeated runs of a heuristic (in this case, TB). For example, the (local minima) solutions resulting from separate TB runs may have differing objective function values, as well as different supply points in the solution set. HC takes advantage of the fact that there is frequently a great deal of overlap in the solution sets corresponding to these local minima. First, HC builds a *concentration set* (CS) by taking the union of the several local minima solutions. The CS has a high likelihood of containing the supply points that make up the optimal solution set. The second stage locates the best solution from among the members of the CS; if the size of the CS is sufficiently small, an integer linear program can be used to optimally select the best solution from the CS.

6.3 A Decomposition Heuristic for DPM2

The heuristics for Discrete-PM2 we presented in the previous section have high computational complexity: recall that a single major iteration of TB takes $O(pn^3)$ time, whereas the HC metaheuristic requires as input the solutions returned by several independent runs of TB. This level of computational effort may be acceptable for typical facility location problems involving between a few hundred to a few thousand demand points, especially when the solutions can be obtained off-line. However, these heuristics do not scale well to problem instances from the application domain we are considering, i.e., when the size of the demand set is equal to the user population served by the network provider, which can be in the hundreds of thousands or more. Hence, we are compelled to design a new heuristic algorithm for DPM2 that can tackle efficiently very large instances of the problem without sacrificing much in solution quality.

We now present an efficient heuristic algorithm for DPM2 that produces good quality results for a variety of input demand distributions. The new algorithm uses as building blocks the vertex substitution heuristic of Teitz and Bart (TB) [111] and the heuristic concentration (HC) approach of Rosing and ReVelle [100] that we described in the previous section. The heuristic uses a decomposition approach to build the concentration set (CS), hence we refer to it as the "decomposition heuristic" (DH). Specifically, DH considers the two DPM1 problems that arise by projecting the n original demand points on the x and y axis, respectively. It then solves the two problems independently using the $O(pn)$ dynamic programming algorithm we described in Chapter 3, and combines the two solutions into a set of points in the plane that make up the CS. The motivation for following this decomposition approach is that, if the x and y coordinates of the demand points in the original DPM2 instance are not correlated (i.e., selected independently), then the CS constructed in this manner is likely to include points that either are in the optimal solution set Z for DPM2 or are close to supply points in set Z.

The DH algorithm follows the two-stage approach of heuristic concentration; a pseudocode description of the algorithm is provided in Fig. 6.3. In the first stage, DH builds the concentration set by running the optimal algorithm for DPM1 twice, once on the x-values and once on the y-values of the n demand points (refer to Steps 1-4 in Fig. 6.3). As a result the algorithm obtains two sets of p points on the x and y axis, respectively; these points correspond to the p best x's and the p best y's, respectively, *when each dimension is considered independently of the other.* Crossing these two sets (Step 5 of the algorithm) yields p^2 points that form the CS. In the second stage, DH randomly generates a number k of initial solutions from among all possible candidate points[1]. Each initial solution serves as input into a separate run of the TB heuristic, which only chooses points for exchange from the CS. Finally, the algorithm returns the best solution among the ones computed by the k TB runs. Note that the final solution returned by DH may include a point that is *not* in the CS. Although TB is restricted to consider only members of the CS for possible replacement points, each initial solution is drawn from *all* candidate points; hence, a solution produced by TB may include points outside the CS if attempting to replace such points in the initial solution did not improve the objective function.

The running time of the first stage of DH is the sum of the time required to solve DPM1, $O(pn)$, and to form the product of the two resulting sets, $O(p^2)$. Since $p < n$, the CS of size p^2 is built in time $O(pn)$. A major iteration of TB considers all p points of a solution, and for each it examines the p^2 points in the CS as possible replacements by evaluating the distance between each and the n demand points. Hence, each major iteration of TB in this case runs in time or $O(np^3)$. In contrast, each major iteration of the original TB runs in time $O(n^3p)$. Thus, the overall complexity of DH represents a substantial improvement over TB, especially for applications, as the one we are considering, in which $n \gg p$. Therefore, DH can be applied to problem instances with very large values of n.

Finally, we note that the DH algorithm can be easily extended to solve the directional p-median problem in $d > 2$ dimensions; such a problem corresponds to tiering structures involving bundles of more than two services.

---

[1] Recall from Chapter 2 that the set of candidate points includes all the demand points as well as all the directional intersection points.

Decomposition Heuristic (DH) for DPM2
Input: A demand set $X = \{(x_1, y_1), \ldots, (x_n, y_n)\}$, the set of candidate points C, and the number of supply points p.
Output: A set Z of supply points.

begin
    // Stage 1: use decomposition to generate the concentration set (CS)
1. $X' \leftarrow \{x_1, \ldots, x_n\}$ // set of x-coordinates of the demand points in X
2. $Z' = \{z'_1, \ldots, z'_p\} \leftarrow$ optimal solution to DPM1 with demand set X'
3. $Y' \leftarrow \{y_1, \ldots, y_n\}$ // set of y-coordinates of the demand points in X
4. $T' = \{t'_1, \ldots, t'_p\} \leftarrow$ optimal solution to DPM1 with demand set Y'
5. $CS \leftarrow Z' \times T'$
    // Stage 2: apply multiple runs of TB to solve DPM2
6. Generate k random initial solution sets $S_i \subset C, i = 1, \ldots, k$, from the candidate points
7. **for** $i \leftarrow 1$ **to** k **do**
8. $Z_i \leftarrow$ TB solution on input S_i // use points for exchange from the CS only
9. **endfor**
10. $Z \leftarrow$ best solution among $\{Z_i\}$
11. **return** Z
end

Fig. 6.3 The decomposition heuristic for DPM2.

6.3.1 Evaluation of the Decomposition Heuristic

We now present the results of an experimental study to evaluate the performance of DH algorithm. Our objective is two-fold. First, we are interested in quantifying the impact of tiered service, in terms of the additional resources needed to map the original demand points to the supply points. Second, we intend to evaluate the quality of the solutions produced by DH relative to those obtained using the original version of the TB heuristic, which is known to attain high-quality results across a broad spectrum of problem instances. Recall that the leftmost term of the right hand side of the DPM2 objective function (6.3) represents the cost of a solution, i.e., the cost of mapping the demand points to a given set of feasible supply points. In order to compare the performance of the DH and TB algorithms across problem instances generated from various demand distributions, we normalize this cost with respect to the total demand ρ_X of the input demand set X. This *normalized DPM2 cost*, expressed as: [2]

---

[2] When the two dimensions of the problem correspond to qualitatively different parameters (e.g., access speed and online storage, respectively), the expression (6.4) for the normalized cost must be modified slightly to ensure that all terms are expressed in the same units. Specifically, the numerator of (6.4) must be written as $\sum_{j=1}^{p} n_j(az_j + bt_j)$, where a and b are appropriate weights to transform the cost of the two parameters to the same units, e.g., US\$. Similarly, the denominator of (6.4) should be changed to $\rho_X = a\sum_{i=1}^{n} x_i + b\sum_{i=1}^{n} y_i$. For simplicity, and without loss of generality, all the results we present in this section were obtained for $a = b = 1$.

$$\text{Normalized DPM2 cost} \quad = \quad \frac{\sum_{j=1}^{p} n_j(z_j + t_j)}{\rho_X} \geq 1 \qquad (6.4)$$

is analogous to the normalized bandwidth requirement (NBR) we defined for the DPM1 and its variants, and is the performance measure we use in this study. Clearly, the closer this quantity is to one, the better the performance of the algorithm, and the smaller the resource penalty to the network due to tiered service.

In our simulation study we run the DH and TB algorithms on a variety of problem instances for which the demand sets X were generated stochastically. Specifically, for each problem instance we generated random points in the plane as follows. For each random point, its x-coordinate was first determined by drawing a value from a given probability density function (PDF); then, its y-coordinate was determined separately and independently by drawing from another (possibly the same) PDF. We considered three different demand distributions, defined in Table 6.1: Uniform (U), Bimodal (B), and Quadrimodal (Q). An input combination is denoted by a two-letter combination, UU, UB, BB, or QQ, in which the first (respectively, second) letter represents the PDF used to generate the x (respectively, y) coordinate of the demand points. Fig. 6.4 shows scatter plots of demand sets of size $n = 1000$ generated from the four input combinations.

Figs. 6.5-6.8 present results corresponding to the four input combinations, UU, UB, BB, and QQ, respectively, and problem instances of size $n = 100, 200$. Each demand set was generated starting from a unique seed for a Lehmer random number generator [72] with modulus $2^{31} - 1$ and multiplier 48271. Each demand set then served as input to the TB and DH heuristics. The TB result for each demand set is the best of 100 independent runs of the TB heuristic. Similarly, the DH result for each demand set is the best among $k = 100$ TB runs during the second stage of the heuristic (refer also to Steps 6-10 of the algorithm in Fig. 6.3).

The graphs in each figure plot the normalized DPM2 cost measure given by expression (6.4) against the number p of supply points. For demand sets of size $n = 100$, we calculated the normalized cost for $p = 5, 10, 15, 20, 25, 30$. For demand sets of size $n = 200$, we calculated the normalized cost for $p = 5, 10, 15, 20, 25$, except for QQ demand sets for which we also calculated the normalized cost for $p = 30$. Each point plotted is the mean of 50 problem instances, with error bars designating a 95% confidence interval.

Each of Figs. 6.5-6.8 contains two pairs of curves, corresponding to demand sets of size $n = 100$ and $n = 200$, respectively. Within each pair, one curve corresponds to the results obtained by the TB heuristic, and the other to results returned by DH.

Comparing the results shown in the four figures, we notice that, although the absolute values do depend on the input combination, the overall relative behavior of the curves is similar across the four input combinations. In particular, we can make two broad observations. First, we note that, for each input, the pair of curves corresponding to $n = 200$ curve lie above the pair corresponding to $n = 100$. Holding p fixed, and for a given algorithm, it is reasonable to expect a higher penalty for a demand set of size $n = 200$ as compared to one of size $n = 100$. There are two factors contributing to this higher penalty. For one, there is a higher "quantization error"

Table 6.1 Formulae for the PDF $f(x)$ of the three demand distributions used to generate DPM2 problem instances.

Distribution	$f(x)$	Domain
Uniform	1/1000	[1, 1000]
Bimodal	1/4000	[1, 250], [351, 650], [751, 1000]
	16/4000	[251, 350], [651, 750]
Quadrimodal	1/4800	[1, 95], [106, 145], [156, 445],
		[456, 595], [606, 1000]
	96/4800	[96, 105], [146, 155],
		[446, 455], [596, 605]

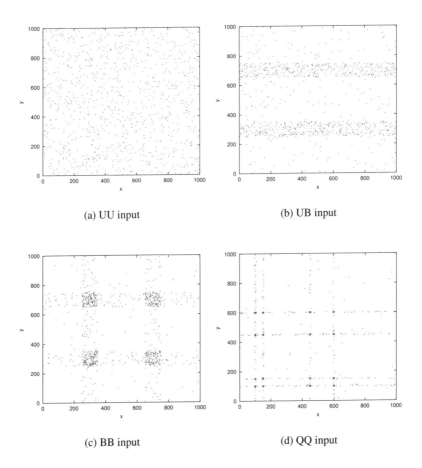

(a) UU input

(b) UB input

(c) BB input

(d) QQ input

Fig. 6.4 Scatter plots of $n = 1000$ for four combinations of the demand distributions shown in Table 6.1.

introduced due to the fact that a larger set of demand points is represented by the same number of supply points; this error is present even in the optimal solution, as the results for the optimal DPM1 algorithm in Chapter 3 demonstrated. Additionally, in this case there is an additional error introduced due to the heuristic nature of the algorithms. Note that the size of the state space that needs to be explored by the DPM2 algorithm increases significantly as the problem size increases from $n = 100$ to $n = 200$. Since the number of runs executed by the two algorithms were the same for both problem sizes, they consider the same number of possible solutions in both cases. In other words, the fraction of the state space explored by the algorithms is much smaller when $n = 200$, resulting in higher error.

We also note that, for a given input and a given problem size, the DH curve lies slightly above the TB curve. In fact, the DH produces solutions whose cost, on average, is less than 5% above the cost of the corresponding TB solutions, across the range of input distributions we considered in this study. In other words, we pay little penalty in solution quality for using DH instead of TB, yet we realize great gains in speed.

In Fig. 6.9 we present results of the DH heuristic on instances of size $n = 1000$. Since the number of candidate points is $O(n^2)$, then moving from an instance of size $n = 100$ to one with $n = 1000$ represents a 100-fold increase in problem size. Due to the high complexity of the TB heuristic, we were not able to obtain results for $n > 200$ within a reasonable amount of time, (e.g., a few hours). Each point in the graphs is the mean of 25 problem instances, and for each instance, DH takes the best solution of 10 runs (i.e., $k = 10$ is Step 6 of the algorithm in Fig. 6.3); error bars along the curves designate a 95% confidence interval. We observe that the curves corresponding to the four input combinations (UU, UB, BB, and QQ) exhibit the same relative behavior we described above. We also note that the curve for a particular input combination lies just above the corresponding curve for $n = 200$ in Figs. 6.5-6.8, indicating that the incremental penalty as the problem size increases is relatively small.

We also note that, for all values of p, the curves for input UU are the highest, followed by UB, BB, and QQ, in this order[3]. Similar observations can be made by comparing the corresponding curves in Figs. 6.5-6.8. We can attribute this result to the nature of the input. Namely, in the sample scatter plots for each input shown in Fig. 6.4, the level of "order" increases as we move from UU to UB to BB to QQ. (Or, said another way, the level of randomness decreases as we move from UU to UB to BB to QQ.) The UU input appears to be the worst-case scenario, with points scattered evenly over the plane (refer to Fig. 6.4). In contrast, the other inputs possess natural clusters where points fall with higher density: UB has two horizontal strips, BB has four squares, and QQ has 16 squares. The more natural

---

[3] There is one exception: the QQ value for $p = 10$ lies above the corresponding BB value. This can be explained by noting that, as Fig. 6.4 indicates, the majority of QQ demands are concentrated in 16 clusters. Hence, $p = 10$ supply points are not sufficient to cover adequately these clusters and the cost of the QQ input is high. However, as p increases to 20, the algorithm has more flexibility in locating the supply points close to the demand points, and the cost of the QQ input drops below the BB cost.

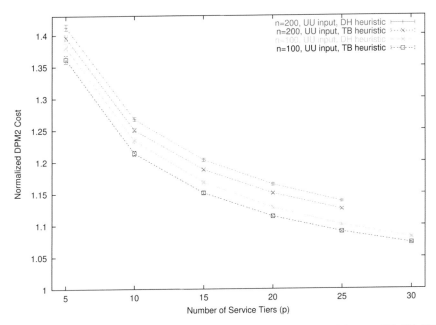

Fig. 6.5 Normalized DPM2 cost vs. p, TB and DH heuristics, instances of size $n = 100, 200$, UU input.

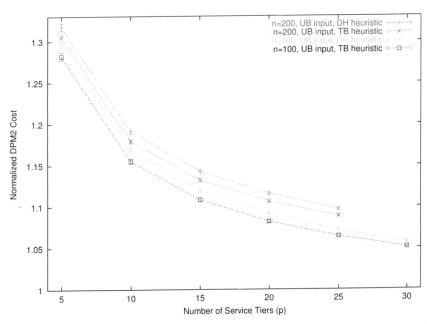

Fig. 6.6 Normalized DPM2 cost vs. p, TB and DH heuristics, instances of size $n = 100, 200$, UB input.

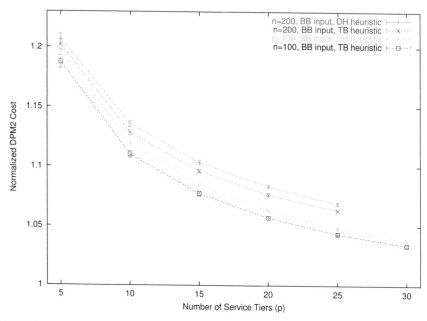

Fig. 6.7 Normalized DPM2 cost vs. p, TB and DH heuristics, instances of size $n = 100, 200$, BB input.

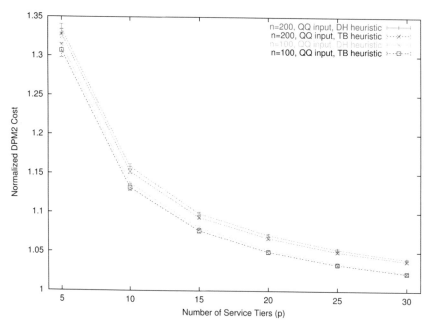

Fig. 6.8 Normalized DPM2 cost vs. p, TB and DH heuristics, instances of size $n = 100, 200$, QQ input.

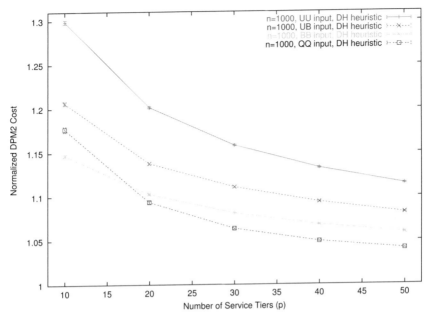

Fig. 6.9 Normalized DPM2 cost vs. p, DH heuristic, instances of size $n = 1000$.

clustering that exists in the input, the lower the penalty due to tiered-service, as the DPM2 algorithm can better align the tiered structure to the set of user demands by locating the supply points close to the clusters of demand points.

Finally, let us compare the results of the DH heuristic for the DPM2 problem in Figs. 6.5-6.9 to the results of the optimal algorithm for the DPM1 problem in Figs. 3.3-3.8, in terms of the normalized cost measure. We find that, in the two-dimensional case, we need a larger number p of supply points in order to keep the normalized cost below a given value. This result is expected, since the state space of the 2-dimensional problem is the product of the corresponding 1-dimensional problems, and intuitively, one would need more supply points to represent the demand points with the same accuracy. However, we note that the value of p required to achieve a certain normalized cost value in two dimensions is significantly *less* than the square of the value of p required to achieve the same value in the 1-dimensional problem, indicating the increasing benefits achievable by employing tiering structures in multiple dimensions.

Overall, the results presented in this section indicate that, compared to the robust TB heuristic, the DH algorithm enjoys substantial speedup while sacrificing little in terms of solution quality, making it an ideal choice for tiered-service network applications with demand set sizes in the hundreds of thousands or larger.

6.4 The Class of Strictly Dominating Solutions for DPM2

Consider a feasible set $Z = \{(z_1,t_1),\ldots,(z_p,t_p)\}$ of supply points for an instance of DPM2. Let (z_i,t_i) and (z_j,t_j) be two supply points in Z. We say that supply point (z_i,t_i) *x-dominates* (respectively, *y-dominates*) supply point (z_j,t_j) if $z_i > z_j$ (respectively, $t_i > t_j$). We say that supply point (z_i,t_i) *strictly dominates* supply point (z_j,t_j), denoted by $(z_i,t_i) \succ (z_j,t_j)$, if:

$$z_i > z_j \quad \text{and} \quad t_i > t_j. \tag{6.5}$$

In other words, one supply point strictly dominates another if the values of the coordinates of the former exceed those of the latter in both dimensions.

We will say that a feasible set Z is a *strictly dominating set* if the p supply points in Z can be labeled so that:

$$(z_p,t_p) \succ (z_{p-1},t_{p-1}) \succ \ldots \succ (z_1,p_1), \tag{6.6}$$

that is, the supply points can be arranged in an order such that each point strictly dominates the previous one in the order. The feasible solution sets obeying expression (6.6) are said to belong to the class of *strictly dominating solutions*. The above definitions generalize to more than two dimensions in a straightforward manner. Fig. 6.10 illustrates a strictly dominating set of 4 supply points for a sample instance of DPM2.

In general, the optimal solution to DPM2 may not belong to the class of solutions defined in (6.6). However, a strictly dominating solution may be of high practical value in the context of tiered network service. Note that each tier is associated with a price (e.g., monthly fee) that the operator charges for offering the services. Typical tiered pricing structures impose a strict ordering of the tiers with respect to price, and the provider's objective is to increase revenue by enticing users to upgrade to a more expensive tier. However, users may perceive that they receive higher utility (value) by upgrading to a more expensive tier that corresponds to a higher level for both services, compared to one that increases the level of one service but decreases (or keeps the same) the level of the other service. Therefore, it might make sense for the provider to offer tiered structures that fall within the strictly dominating class.

We now show that the optimal solution within the class of strictly dominating solutions can be found in polynomial time using dynamic programming. Let $\hat{X} = \{x_1,\ldots,x_l\}$ be a set containing the l distinct x-coordinates of the n demand points in X, $l \leq n$, and assume that the elements of the set are labeled such that $x_1 < x_2 < \ldots < x_l$. Also let $\hat{X}_i = \{x_1,\ldots,x_i\}, i = 1,\ldots,l$, denote the subset of $\hat{X}$ containing the i smallest elements. Similarly, define $\hat{Y} = \{y_1,\ldots,y_m\}, m \leq n$, as the set with m distinct y-coordinates, labeled in increasing order, of the demand points in X; sets $\hat{Y}_j = \{y_1,\ldots,y_j\}, j = 1,\ldots,m$, are defined in a manner analogous to sets $\hat{X}_i$. Finally, let $n(\hat{X}_i,\hat{Y}_j)$ denote the number of demand points whose x- and y-coordinates are no greater than x_i and y_j, respectively.

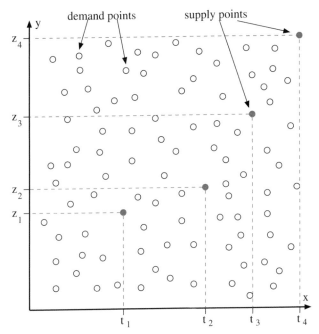

Fig. 6.10 A strictly dominating set of 4 supply points for a sample instance of DPM2

Define $\Xi(\hat{X},\hat{Y},p)$ as the value of the left term in the expression (6.3) of the DPM2 objective function for the optimal solution within the class of strictly dominating solutions, when the number of supply points is p. Since the right term ρ_X of (6.3) is constant given the demand set X, minimizing $\Xi(\hat{X},\hat{Y},p)$ is equivalent to minimizing the DPM2 objective function. Within the class of strictly dominating solutions, the optimal value of $\Xi(\hat{X},\hat{Y},p)$ can be obtained through the following dynamic programming algorithm:

$$\Xi(\hat{X}_1,\hat{Y}_1,r) = x_1 + y_1, \quad r = 1,\ldots,p \tag{6.7}$$

$$\Xi(\hat{X}_i,\hat{Y}_j,1) = n(\hat{X}_i,\hat{Y}_j) \times (x_i + y_j), \quad i = 1,\ldots,l, \ j = 1,\ldots,m \tag{6.8}$$

$$\Xi(\hat{X}_i,\hat{Y}_j,r+1) = \min_{\substack{h=r,\ldots,i-1 \\ k=r,\ldots,j-1}} \left\{ \Xi(\hat{X}_h,\hat{Y}_k,r) + \left[n(\hat{X}_i,\hat{Y}_j) - n(\hat{X}_h,\hat{Y}_k)\right] \times (x_i + y_j) \right\}$$

$$r = 1,\cdots,p-1; \ i = 2,\cdots,l; \ j = 2,\cdots,m \tag{6.9}$$

Expression (6.7) simply states that if there is a only one demand point, it is the optimal supply point. Expression (6.8) is due to the fact that, if there is only a single supply point, then it is the demand or intersection point with the largest x and y co-ordinates, i.e., point (x_i, y_j); moreover, all demand points are mapped to this supply point, and there are $n(\hat{X}_h,\hat{Y}_k)$ demand points in the given sets. Finally, the recursive expression (6.9) can be explained by noting that the $(r+1)$-th supply point must be

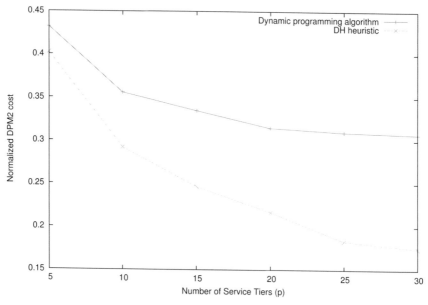

Fig. 6.11 Normalized DPM2 cost vs. p, $n = 100$, UU input.

equal to the demand or intersection point (x_i, y_j). If the r-th supply point coincides
with the demand or intersection point $(x_h, y_k), h < i, k < j$, then all demand points
with coordinates greater than x_h or y_k, respectively, are mapped to point (x_i, y_j);
there are $\left[n(\hat{X}_i, \hat{Y}_j) - n(\hat{X}_h, \hat{Y}_k)\right]$ such points. Taking the minimum over all possible
values of h and k provides the optimal value. A straightforward implementation
of the above dynamic programming algorithm has a running time complexity of
$O(pn^3)$.

Fig. 6.11 plots the normalized DPM2 cost of solutions obtained by the the dynamic programming and DH algorithms for demand sets of size $n = 100$ and the
UU distribution, as a function of the number p of supply points. We observe that
the cost of the solutions produced by the dynamic programming algorithm (6.7)-
(6.9) is higher than that of the solutions found by the DH algorithm. The relative
difference in cost also increases with p. This can be explained by noting that the
space of solutions explored by the dynamic programming algorithm is a small subset of the solution space explored by DH. More specifically, an algorithm that can
select any candidate point as a supply point will be able to take advantage of a
larger value for p to better align the supply points with the demand points. On the
other hand, an algorithm that is limited to producing strictly dominating solutions is
inherently limited in its options with respect to locating the supply points. As a result, additional supply points produce little benefit, hence the curve of the dynamic
programming algorithm in Fig. 6.11 flattens out quickly. Nevertheless, the dynamic
programming algorithm is optimal whenever the set of supply points has to belong

the class of strictly dominating solutions, and it also performs well for general problem instances with a small number p of supply points, as is the case in the network context we are considering.

Part II
Economics

Chapter 7
Economic Model for Bandwidth Tiered Service

In Part I of the book, we considered several variants of the problem of determining optimal structures for tiered services. These problem formulations reflect cost considerations on the part of the network provider only: indeed, the various objective functions represent measures of the additional cost incurred by operating a tiered-service network compared to a continuous-rate one, and the goal is to minimize this cost. In Part II, on the other hand, we use concepts from economic theory to formulate the service tier selection problem in a manner that takes into account the realities of the marketplace, and in particular, the interests of users, in addition to the cost incurred by the provider for offering the service. Our work provides a theoretical framework for reasoning about and pricing Internet tiered services, as well as a practical toolset for network providers to develop customized menus of service offerings. Our results also indicate that tiering solutions currently adopted by ISPs perform poorly both for the providers and the society overall.

In this chapter we consider a network provider offering a service characterized by a single parameter. Without loss of generality, in the remainder of the chapter our discussion will focus on a bandwidth tiered service, e.g., as it applies to broadband Internet access; however, all our formulations and results can be readily adapted to other types of network services. To capture the interests of users, we assume that the value that a user receives from the service can be expressed as a function of the service itself. We only consider a homogeneous user population, in that all users are assumed to receive the same value from the service (or, put another way, all users are characterized by the same value or utility function); in the following chapter we will remove this assumption and consider heterogeneous user populations. We develop an economic model for such networks and make contributions in two important areas. First, we formulate the problem of selecting the service tiers from three perspectives: one that considers the users' interests only, one that considers only the service provider's interests, and one that considers both simultaneously, i.e., the interests of society as a whole. We also present an approximate yet accurate and efficient solution approach for tackling these nonlinear programming problems. Given the set of (near-) optimal service tiers, we then employ game-theoretic tech-

G.N. Rouskas, *Internet Tiered Services*, DOI: 10.1007/978-0-387-09738-1_7,
© Springer Science + Business Media, LLC 2009

niques to find an optimal price for each service tier that strikes a balance between the conflicting objectives of users and service provider.

7.1 Pricing of Internet Services

Multi-tiered price systems are prevalent for both business and residential Internet access, and have been employed in various forms regardless of whether the underlying pricing scheme is capacity-based (in which the subscription fee is determined solely by the user's access speed) or usage-sensitive (in which price is a function of the actual bytes transferred over a certain time period, usually one month). The introduction of tiered service has important engineering and financial implications for the network provider. On the engineering front, tiered-service models have the potential to simplify a wide range of core functions related to the operation, control, and management of the network, and hence provide significant advantages in terms of scalability. From a financial standpoint, multitiered pricing schemes, if designed and applied appropriately, can be a catalyst for Internet service innovation and penetration. Tiered structures can be an effective tool for ISPs to optimize and specialize their offerings so as to capitalize on the increasing sophistication and requirements of various segments of Internet users, as well as to differentiate themselves from the competition. Tiered pricing also has the potential to spur user adoption by providing a wide menu of customized services from which users may select based on needs and affordability.

To realize this potential, it is crucial that both the service tiers and the corresponding prices be determined in a manner that takes into account simultaneously the (usually conflicting) objectives of users and providers. In current practice, however, there is considerable lack of transparency in how ISPs set their tiered price structures, and it is unclear whether the perspective of users is even considered in the process. For instance, certain service tiers for business Internet access are based on the bandwidth hierarchy of the underlying network infrastructure (e.g., DS-1, OC-3, etc.). While this is a natural arrangement for the service provider, it is unlikely that hierarchical rates designed decades ago for voice traffic would be a good match for today's business data applications. For residential Internet access, on the other hand, ISPs have adopted simple exponential tiering structures, both for capacity-based pricing (e.g., as is the case for the ADSL tiers available from various providers) and for usage-sensitive pricing [50, 51]. The relationship between these exponentially increasing levels of service (and their price) and the usage patterns (and willingness or ability to pay) of the population of potential subscribers is open to debate.

Pricing of Internet services using concepts from economic theory has been a subject of research for more than a decade [47, 62–64, 76, 84, 86, 97, 107]. This is a broad area that encompasses issues from calculating the cost of resources to determining the services to offer and setting appropriate prices, and from dealing with the realities and economics of layered networks to interconnection agreements between ISPs. An initial focus was on charging as a mechanism for controlling the behavior of users, and/or for limiting usage to make room for higher paying users. Early work [63, 64] also addressed the issues of charging, rate control, and routing in communication networks carrying elastic traffic. The main finding of these studies was that the system reaches an optimum state when the network's choice of allocated rates is at equilibrium with users' choices of charges. The Paris Metro Pricing (PMP) scheme in [86] separates the network into independent subnetworks that behave similarly but charge their customers at different rates. A mathematical model of PMP was developed in [97] by viewing each subnetwork as a single bottleneck queue, and assuming that data packets may select the most suitable subnetwork "intelligently" by considering not only the delays but also the prices charged. The conclusion of the study was that there exist necessary and sufficient conditions for the system to attain stability. The issue of charging at the session or network layer while maintaining a clean separation between the underlying technologies was considered in [62].

More recent work has studied the issues arising in pricing multiple classes of service, especially in the context of differentiated services. A game theoretic pricing mechanism for "statistically guaranteed" service in the Internet was proposed in [107]. This mechanism was shown to offer better service and lower prices to users, and enables the provider to adopt various service and revenue models. The work in [47] also adopted a game-theoretic strategy to study a simple two-class differentiated service model, and found that the system is easy to trap into an undesirable equilibrium whenever prices do not properly reflect the quality of the service provided. Accordingly, a new dynamic pricing approach was proposed in order to avoid this problem. Finally, a free market economic model for ad-hoc wireless networks was proposed recently in [84]. Based on a greedy pricing strategy, the model maximizes the social welfare while ensuring non-negative profit for the users and service provider. This study also developed a non-greedy policy that optimizes a profit fairness metric.

The work we present in this chapter differs from existing literature in that our focus is on optimizing the service tiers and corresponding price structures given some information about users and providers, regardless of the underlying assumptions upon which this information is based. Consequently, this work is quite general in scope and may be applied to a variety of contexts, independently of whether the pricing scheme is capacity-based or usage-sensitive, whether charging is at the network or session/application layers, or whether the transaction is between users and provider or between providers.

7.2 The Network Context

We consider a network that offers a service characterized by a single parameter, e.g., the bandwidth of the user's access link, and charges users on the basis of the amount of service they receive. Users may request any amount of service depending on their needs and their willingness or ability to pay the corresponding service fee. We adopt a stochastic model for the user demands, similar to the one we discussed in Chapter 5. Specifically, we assume that the distribution of the size x of user service requests is known; such a distribution may be obtained empirically, or extrapolated from observed user behavior and application requirements. Let $f(x)$ and $F(x)$ be the probability density function (PDF) and cumulative distribution function (CDF), respectively, representing the population of user requests. The PDF and CDF are defined in the interval $[x_{min}, x_{max}]$, where x_{min} and x_{max} correspond to the minimum and maximum, respectively, amount of service requested by any user. In contrast to the model we used in Chapter 5, we do not assume that user demands are normalized in any way; hence the domain $[x_{min}, x_{max}]$ of $f(x)$ and $F(x)$ can be any interval on the real line.

The network offers p levels (tiers) of service, where typically p is a small integer, much smaller than the number n of network users (i.e., $p \ll n$). As in previous chapters, we define $Z = \{z_1, z_2, \ldots, z_p\}$ as the set of distinct service tiers offered by the network provider; without loss of generality, we assume that the service tiers are distinct and are labeled such that $z_1 < z_2 < \ldots < z_p$. For notational convenience, we also define the "null" service tier $z_0 = 0$. With tiered service, a user with service request $x, x_{min} \leq x \leq x_{max}$, subscribes to service tier z_j such that $z_{j-1} < x \leq z_j$, as illustrated in Fig. 5.1 of Chapter 5. Note that Fig. 5.1 represents a special case of the more general scenario we consider in this chapter, since it assumes that $x_{min} = 0$ and $x_{max} = 1$. As we mentioned in Chapter 5, in order to accommodate all user requests, the highest tier must be such that $z_p = x_{max}$, an assumption we will make throughout this work.

The network provider incurs a cost for the service it provides, and consequently, it will be inclined to select the service tiers, and the corresponding price to charge, so as to recoup its costs (and make a profit). On the other hand, each user subscribes to a service that is at least as good as the one requested, but the additional value, if any, that the user receives may be offset by the higher cost of the service. Our aim is to apply economic theory to capture analytically these tradeoffs, and to develop techniques to select the service tiers and prices in a manner that accounts for both the users' and providers' perspectives.

To develop an economic model for tiered-service networks, we assume the existence of three non-decreasing functions of service x, as shown in Fig. 7.1. The *utility* function, $U(x)$, is a measure of the value that users receive from the service, and it stands for their willingness to pay for the service. The *cost* function, $C(x)$, represents the cost incurred by the provider for offering the service. Finally, the *price* function, $P(x)$, represents the amount that the service provider charges for the service. Fig. 7.1 shows that $U(x)$ lies above $P(x)$ (otherwise users would not be willing to pay for the service), and in turn $P(x)$ lies above $C(x)$ (otherwise providers would not be inclined to offer the service); however, all the results in this chapter are valid for general functions $U(x)$, $P(x)$, and $C(x)$, independently of the relative behavior of the corresponding curves. We make the reasonable assumption that utility, cost, and price are all expressed in the same units (e.g., US$).

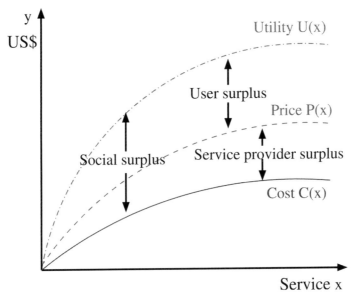

Fig. 7.1 Utility, cost, and price functions (© 2008 IEEE)

Before we proceed to present an economic model for optimally selecting the service tiers, we make two important observations. First, we note that utility and cost typically depend only on the user and service provider, respectively, but that price is the result of market dynamics and the relative bargaining power of users and service providers. The practical implication is that the utility and cost functions can be assumed to be known (or estimated) in advance, but the same is not necessarily true for the price function. Second, the cost function $C(x)$ reflects the absolute cost to the operator for providing an amount x of service. This cost model is different than the one we adopted in Part I of the book, as here we are not concerned with cost comparisons to a continuous-rate network.

7.3 Economic Model for Sizing of Service Tiers

In this section, we use concepts from economics to describe the relationship between users and service providers, and we propose optimization problems for selecting the service tiers. We also illustrate how to solve these problems to obtain a set of (near-) optimal tiers. In the following section, we apply Nash bargaining theory [81, 83] to determine optimal pricing strategies for this fixed (near-optimal) set of tiers.

Consider now the demand-supply relationship between the users and network service providers. On the one hand, users want to maximize the utility they obtain from the service while keeping the fee they have to pay to the service provider as low as possible; in economic terms, users want to maximize the *user surplus* [8, 36], defined as the difference between the utility they obtain from the service and the price they have to pay for it. On the other hand, the network providers' objective is to charge a high fee so as to offset the cost of offering the service and make a profit; in other words, service providers want to maximize the *service provider surplus* [8, 36], defined as the difference between price and cost. The concepts of user surplus and service provider surplus are illustrated in Fig. 7.1.

From the point of view of the society as a whole, it is preferable to maximize the overall *social welfare*, defined as the sum of the user surplus plus the provider surplus (see also Fig. 7.1). We will refer to the social welfare as *social surplus* [8, 36]. Once the maximum social surplus has been determined, the users and service providers may negotiate its division into user and service provider surpluses through bargaining, as we explain in the next section.

Let us use $S_{usr}(x)$, $S_{pr}(x)$, and $S_{soc}(x)$ to denote the user, provider, and social surplus functions, respectively. These are defined as follows:

$$S_{usr}(x) = U(x) - P(x) \tag{7.1}$$
$$S_{pr}(x) = P(x) - C(x) \tag{7.2}$$
$$S_{soc}(x) = U(x) - C(x) \tag{7.3}$$

In the tiered-service network under consideration, the problems of maximizing the surplus of users, service providers, or society, amount to selecting appropriately the set of service tiers to be offered, as we discuss next.

7.3.1 Maximization of Expected Surplus

Let $S(x)$ be the surplus function (i.e., one of $S_{usr}(x), S_{pr}(x)$, or $S_{soc}(x)$, defined above), and suppose for the moment that the set $Z = \{z_1, \ldots, z_p = x_{max}\}$ of p service tiers is given. In this case (refer also to Fig. 5.1), all users with requests in the interval[1] $(z_{j-1}, z_j]$ subscribe to tier z_j, incurring a surplus of $S(z_j), j = 1, \ldots, p$. Recalling that $f(x)$ and $F(x)$ are the PDF and CDF, respectively, of user requests, the *expected* surplus $\bar{S}(z_1, \ldots, z_p)$ for the given service tier vector Z can be expressed as:

$$\bar{S}(z_1, \ldots, z_p) = \sum_{j=1}^{p} \left(\int_{z_{j-1}}^{z_j} S(z_j) f(x) dx \right)$$
$$= \sum_{j=1}^{p} \left(S(z_j) \int_{z_{j-1}}^{z_j} f(x) dx \right)$$

---

[1] Note that the leftmost interval is $(z_0, z_1]$, where $z_0 = 0$ is the "null" service tier we defined earlier. Since $F(z_0) = 0$, the summation in expression (7.4) is correctly defined for all service tier intervals.

$$= \sum_{j=1}^{p} \left(S(z_j) \left(F(z_j) - F(z_{j-1}) \right) \right). \tag{7.4}$$

Consider now the problem of optimally selecting the service tiers *from the users' point of view*. Based on our earlier discussion, the objective of each network user is to maximize its surplus. Considering all the users in the network *as a whole*, the objective is to select the set of service tiers so as to maximize the expected aggregate user surplus, i.e., the weighted sum of the individual user surpluses in expression (7.4) with S_{usr} in place of $S(x)$. Similarly, the goal of the service provider is to maximize its expected aggregate surplus, while considering the welfare of the society (i.e., both users and providers), the objective would be to maximize the expected aggregate social surplus. These last two objectives are obtained by using $S_{pr}(x)$ and $S_{soc}(x)$, respectively, in place of $S(x)$ in (7.4).

These three optimization problems can be formally expressed as instances of the following problem which we will refer to as the *Maximization of Expected Surplus (MAX-ES)* problem. Note that the objective function (7.5) is nonlinear with respect to the variables $z_1, \ldots, z_p$, hence MAX-ES is a nonlinear programming problem.

Problem 7.1 (MAX-ES). Given the CDF $F(x)$ describing the population of user requests, an integer p, and a surplus function $S(x)$, find a set of service tiers $Z = \{z_1, \ldots, z_p\}$ that maximizes the following objective function representing the expected surplus:

$$V(z_1, \ldots, z_p) = \sum_{j=1}^{p} \left(S(z_j) \left(F(z_j) - F(z_{j-1}) \right) \right) \tag{7.5}$$

subject to the constraints:

$$x_{min} < z_1 < z_2 < \ldots < z_p = x_{max}. \tag{7.6}$$

Let us assume for the moment that an optimal solution to MAX-ES can be obtained; we will discuss shortly how to find such a solution. Consider the optimal solution obtained by solving the MAX-ES problem from the perspective of users or providers (i.e., by using the user $S_{usr}(x)$ or provider $S_{pr}(x)$ surplus function in place of $S(x)$, respectively). Such a solution is unlikely to be of practical value, for two reasons. First, it assumes that users and service providers may select the service tiers optimally based only on their own interests. In reality, a service tier vector that is optimal for the users may not be acceptable to the service provider, and vice versa. Therefore, it is important to obtain a jointly optimal solution that takes into account the perspectives of both users and service providers. Second, both the user and provider surplus functions assume the existence of a pricing function $P(x)$. As we explained earlier, in general the price function is the result of marketplace dynamics, including negotiation between users and service providers, hence it may not be known in advance.

On the other hand, the social surplus function $S_{soc}(x)$ depends only on the cost and utility functions, which are generally known (or may be estimated accurately) in advance. Therefore, considering the welfare of the society as a whole overcomes the above difficulties since

- it takes into account simultaneously the interests of both users (through the utility function) and providers (through the cost function), and
- allows one to determine the optimal service tier vector without knowledge of the pricing function.

Therefore, for the remainder of this chapter we will consider the MAX-ES problem from the society's point of view only; although, for simplicity we will continue using $S(x)$ as the surplus function, the reader should keep in mind that from now on we assume that $S(x) = S_{soc}(x) = U(x) - C(x)$.

7.3.2 Solution Through Nonlinear Programming

Returning to the formulation (7.5)-(7.6) of the MAX-ES problem, we observe that it is similar to the formulation of the SDPM1 problem given in (5.6)-(5.7). Indeed, the constraints (7.6) of MAX-ES reduce to the constraints (5.7) of SDPM1 by letting $x_{min} = 0$ and $x_{max} = b$, whereas the objective function (7.5) of MAX-ES becomes identical to the one for SDPM1 by letting the surplus function $S(x) = x$ and introducing a constant term (that does not affect the optimization) corresponding to μ in (5.6). Consequently, SDPM1 can be seen as a special case of MAX-ES, except for the fact that the former is a minimization problem while the latter is a maximization problem. Therefore, MAX-ES can be solved using techniques similar to the ones we used in Chapter 5 to solve SDPM1. For completeness, and due to the fact that MAX-ES is a maximization problem with a more general objective function, we present some new results and the solution techniques for MAX-ES in the following subsections.

If the nonlinear objective function (7.5) of the MAX-ES problem is concave, and since the constraints (7.6) are convex, we may use the Karush-Kuhn-Tucker (KKT) conditions to find the global maximum [13]. The following lemma derives sufficient conditions for the function (7.5) to be concave.

Lemma 7.1. *If $S(x)$ and $F(x)$ are continuous and twice differentiable in $[x_{min}, x_{max}]$ and the two conditions*

$$S''(x)[F(x) - F(y)] + 2S'(x)F'(x) + S(x)F''(x) < 0 \tag{7.7}$$

$$-[S'(x)F'(y)]^2 - S(x)F''(y)\left\{S''(x)[F(x) - F(y)] + 2S'(x)F'(x) + S(x)F''(x)\right\} > 0 \tag{7.8}$$

are satisfied for all $x, y \in [x_{min}, x_{max}]$ *with* $y < x$, *then the MAX-ES objective function* V *is concave in the feasible area* $x_{min} < z_1 < \ldots < z_{p-1} < z_p = x_{max}$.

Proof. Define $\omega(x, y) = S(x)(F(x) - F(y))$. We can then rewrite (7.5) as:

$$V(z_1, \ldots, z_p) = \sum_{j=1}^{p} \omega(z_j, z_{j-1}). \tag{7.9}$$

Since the sum of concave functions is also a concave function, a sufficient condition for V to be concave is for ω to be concave in the feasible area $x_{min} < y < x < x_{max}$.

The Hessian of ω is the symmetric matrix

$$\mathbf{H} = \begin{pmatrix} h_{1,1} & h_{1,2} \\ h_{2,1} & h_{2,2} \end{pmatrix}, \tag{7.10}$$

where

$$h_{1,1} = \frac{\partial^2 \omega}{\partial x^2} = S''(x)[F(x) - F(y)] + 2S'(x)F'(x) + S(x)F''(x) \tag{7.11}$$

$$h_{2,2} = \frac{\partial^2 \omega}{\partial y^2} = -S(x)F''(y) \tag{7.12}$$

$$h_{1,2} = h_{2,1} = \frac{\partial^2 \omega}{\partial x \partial y} = -S'(x)F'(y) \tag{7.13}$$

If ω is continuous and has continuous first and second derivatives, then it is concave if its Hessian is negative definite in the feasible area $x, y \in [x_{min}, x_{max}]$ with $y < x$, or:

$$h_{1,1} < 0 \text{ and } h_{1,1}h_{2,2} - h_{1,2}^2 > 0,$$

from which the two conditions (7.7) and (7.8) follow. $\square$

7.3.3 An Efficient Approximate Solution

As we remarked in Chapter 5.3, the objective function (7.19) of MAX-ES will not be concave whenever the surplus function $S(x)$ and CDF $F(x)$ take general forms, e.g., when $F(x)$ is not continuous. Rather than employing approximate techniques, we follow an approach similar to the one we described in Chapter 5.3. Specifically, we create a discrete approximation of the PDF $f(x)$ that we use to obtain an approximate formulation of MAX-ES that asymptotically converges to the formulation in (7.5)-(7.6). We then develop a dynamic programming algorithm that solves this new problem optimally.

We obtain the discrete approximation of the PDF $f(x)$ as we explained in Chapter 5.3 and illustrated in Fig. 5.3. Recall that we partition the interval $[x_{min}, x_{max}]$ into $K > p$ intervals each of length equal to $\frac{x_{max} - x_{min}}{K}$. The right-hand endpoint of the k-th interval is $e_k = x_{min} + \frac{k(x_{max}-x_{min})}{K}$, and we associate with e_k a discrete point mass density P_k given by expression (5.18). The K pairs $\{(e_k, P_k)\}$ form the approximation of $f(x)$, while the K pairs $\{(e_k, F_k)\}$ form the approximation of the CDF $F(x)$, where quantities F_k are given by expression (5.19).

In order to obtain an efficient solution to the MAX-ES problem, we also impose the additional restriction that the p service tiers may only take values from the set $\{e_k\}$ of the interval endpoints. Consequently, our objective is to solve the following discrete version of MAX-ES, which we will refer to as *Approximate-MAX-ES*. Again we note that the Approximate-SDPM1 problem we introduced in Chapter 5.3.1 can be seen as a special case of Approximate-MAX-ES, along the lines of our earlier discussion regarding the relationship between the MAX-ES and SDPM1 problems.

Problem 7.2 (Approximate-MAX-ES). Given the the K-point approximation $\{e_k, P_k\}$ of the PDF of user requests, an integer number $p < K$ of service tiers, and a surplus function $S(x)$, find a set of service tiers $Z = \{z_1, \ldots, z_p\}$ that maximizes the objective function:

$$\bar{V}(Z) = \bar{V}(z_1, \ldots, z_K) = \sum_{j=1}^{p} \left(S(z_j) \left(F_{k_j} - F_{k_{j-1}} \right) \right) \qquad (7.14)$$

subject to the constraints:

$$z_j = e_{k_j} \in \{e_k\}, \quad j = 1, \ldots, p, \ k = 1, \ldots, K \qquad (7.15)$$

$$z_1 < z_2 < \ldots < z_p = x_{max}. \qquad (7.16)$$

The objective function $\bar{V}(Z)$ in (7.14) represents the *approximate expected surplus* for the given service tier set Z. Note that the term $\left(F_{k_j} - F_{k_{j-1}} \right)$ is the fraction of user requests falling in the interval $(z_{j-1}, z_j]$ and are mapped to tier z_j. Therefore, the j-th term of the sum in the right-hand side of (7.14) is the expected surplus incurred by users mapped to the j-th tier under the PDF approximation. Since the PDF approximation approaches the original PDF as $K \to \infty$, an optimal solution to Approximate-MAX-ES for sufficiently large values of K will similarly approach the optimal solution to the original MAX-ES problem.

Define $\Phi(k, l)$ as the optimal value of the objective function (7.14) when the number of intervals is k and the number of service tiers is $l \leq k$. Then, $\Phi(k, l)$ may be computed recursively as follows:

$$\Phi(k, 1) = S(e_k)F_k, \quad k = 1, \ldots, K \qquad (7.17)$$

$$\Phi(k, l+1) = \max_{q=l, \ldots, k-1} \{ \Phi(q, l) + S(e_k)(F_k - F_q) \},$$

$$l = 1, \ldots, p-1; \; k = 2, \ldots, K \qquad (7.18)$$

The above dynamic programming algorithm is a generalization of the algorithm described in expressions (5.23)-(5.24). Expression (7.17) can be explained by observing that if there is only one tier of service, it must coincide with the right-hand endpoint of the k-th (i.e., rightmost) interval. The recursive expression (7.18) simply states that, for $l+1$ service tiers, the largest tier must coincide with the right-hand endpoint of the k-th interval, and the remaining l tiers must be optimally assigned to the endpoints of any feasible interval $q, l \le q \le k-1$.

The running time of the above dynamic programming algorithm to obtain $\Phi(K, p)$ is $O(pK^2)$. Since the time complexity is a function of the number K of intervals, there is a tradeoff between the quality of the solution (which requires a large value for K to obtain a good approximation of the PDF) and running time. We investigate the convergence of the solution as a function of K later in this chapter.

7.3.4 Optimizing the Number of Service Tiers

The MAX-ES problem takes the number p of service tiers as input, and its objective function (7.5) is based on the assumption that there is no cost associated with offering each tier. In general, however, the total cost to the network provider of offering p tiers of service consists of two components. The first component is due to the cost of bandwidth: the higher the access speed or the amount of traffic generated by the users, the higher the cost. We used the nondecreasing function $C(x)$ to denote this cost, which may be used to represent the link cost for carrying user traffic as well as the cost of switching the traffic in the network. The second component is due to the cost of software and hardware mechanisms (e.g., queueing structures, policing mechanisms, control plane support, etc.) required inside the network for implementing a given number p of service tiers.

Let $C_t(l)$ be a nondecreasing function representing the cost of employing l service tiers. In this variant of the MAX-ES problem, the objective is to determine both the optimal number p of service tiers and their values so as to maximize the objective function:

$$V(z_1, \ldots, z_p) - C_t(p) \qquad (7.19)$$

where $V(z_1, \ldots, z_p)$ is the expected surplus in expression (7.5). The approximate formulation of this variant, along the lines of the one we described for MAX-ES in the previous subsection, is a generalization of the JDPM1 problem we introduced in Chapter 3.4. This approximate problem can be solved optimally with the dynamic programming algorithm of the previous subsection, after modifying the expressions (7.17)-(7.18) to account for the cost component $C_t(k)$, similar to how the expressions (3.22)-(3.24) reflect the per tier cost. The running time of this algorithm is $O(K^3)$, since it has to examine all K possible values for the number of service tiers.

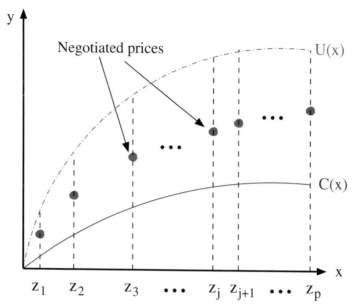

Fig. 7.2 Optimal price vector

After solving the MAX-ES problem, we obtain a service tier set $Z^\star = \{z_1,\ldots,z_p\}$ that maximizes the social surplus and depends only on the utility and cost functions provided by the users and network provider, respectively. Next, we describe an approach to obtaining the optimal price for each service tier in $Z^\star$ in a manner that strikes a balance between the conflicting objectives of users and providers.

7.4 Optimal Pricing Based on Nash Bargaining

Consider a set $Z^\star = \{z_1, z_2, \ldots, z_p\}$ of service tiers that maximizes the social surplus. We are interested in finding an appropriate price $P(z_j)$ for each service tier $z_j, j = 1,\ldots,p$, so as to satisfy both the users and service provider. Clearly, the price for each service tier z_j should be between the values of the cost and utility functions at service level z_j, as illustrated in Fig. 7.2.

In a free telecommunication market, the price for the service is typically the result of a negotiation process between the users and service providers. This negotiation, or bargaining, process can be thought of as a game during which each party attempts to maximize its own surplus [81–83]; the outcome of the game is an optimal price for the service that is mutually acceptable by both parties. This game can be seen as an abstraction of market dynamics, e.g., reflecting the users' ability to compare prices from various providers and competition among providers. We also emphasize that our focus is on a bargaining game that takes place once, after

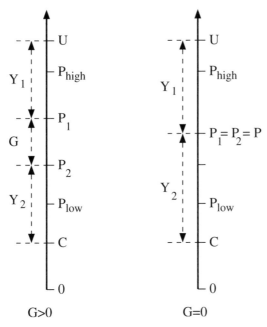

Fig. 7.3 Relationship among the game parameters U, C, G, P_1, P_2, P_{high}, P_{low}, Y_1, and Y_2

which optimal prices are determined and fixed for a relatively long time compared to typical flow durations. In other words, we do not consider the scenario in which the users and/or the network attempt to set prices on a per-session basis (e.g., as in [62–64]), a scenario that we do not believe is practical.

7.4.1 The Single Tier Case

Let us first consider the case $p = 1$, whereby the provider offers a single service tier z_1. For notational convenience, in this section we will simply use U, P, and C, instead of $U(z_1)$, $P(z_1)$, and $C(z_1)$, respectively.

Let P_{high}, $P_{high} \leq U$ be the maximum price that the users would accept as a satisfactory outcome of the negotiating process (game). Similarly, let P_{low}, $P_{low} \geq C$, be the minimum price that the service provider would find acceptable. We use P_1 to denote the price that users pay for the service, and P_2 the amount that the provider receives for the service. In general, there may exist a gap G between the P_1 and P_2; this gap is referred to as *transaction cost* in economics. Without loss of generality, in this work we assume that the gap G has a fixed value; clearly, if $G = 0$, then $P_1 = P_2$. We also define $Y_1 = U - P_1$, and $Y_2 = P_2 - C$. Fig. 7.3 illustrates the relationship among the game parameters U, P_{high}, P_{low}, C, P_1, P_2 and G.

The two parties, users and service provider, are interested in dividing the *net social surplus*, i.e., the social surplus minus the transaction cost, which from Fig. 7.3 is equal to $(U - C - G)$. As we can see, the net social surplus is the sum of Y_1 and Y_2. Y_1 and Y_2 represent the shares of the good to be divided and stand for the excess utility (or net surplus) of users and provider, respectively. The objective is to find an optimal division of the net social surplus such that both parties feel satisfied. This optimization problem was introduced by Nash [81, 83] as a cooperative bargaining game, and is widely used in the literature for characterizing labor negotiations and a range of other bargaining situations [94].

Let $\beta, 0 \leq \beta \leq 1$, be the bargaining power of the users, and $1 - \beta$ be the bargaining power of the service provider. Bargaining power, as defined here, refers to the relative ability of each party in the bargaining game to influence the setting of prices. Then, $\Omega = Y_1^{\beta} Y_2^{1-\beta}$ is the Nash product [81, 83] in the bargain. In essence, Ω is the product of the players' excess utilities, each scaled by the corresponding player's bargaining power. Our objective is to find suitable values for Y_1 and Y_2 that maximize Ω. This *price optimization* (OPT-P) problem can be formulated as:

Problem 7.3 (OPT-P). Given the user utility U, the provider cost C, the highest price P_{high} that users are willing to pay, the lowest price P_{low} the provider is willing to accept, the transaction cost G, and the bargaining power of users $\beta, 0 \leq \beta \leq 1$, and provider $1 - \beta$, maximize the objective function:

$$\max_{Y_1, Y_2} \quad \Omega = Y_1^{\beta} Y_2^{1-\beta} \tag{7.20}$$

subject to the constraints:

$$Y_1 + Y_2 \leq U - C - G \tag{7.21}$$

$$Y_1 \geq U - P_{high} \tag{7.22}$$

$$Y_2 \geq P_{low} - C \tag{7.23}$$

where Y_1 and Y_2 represent the net (i.e., after the transaction cost) surplus of users and provider, respectively.

The constraints (7.21)-(7.23) can be explained by referring to Fig. 7.3. Specifically, expression (7.21) states that the aggregate net surplus may not exceed the gross surplus (defined as the difference between utility and cost) minus the transaction cost G. Expression (7.22) captures the fact that the net surplus of users may not be less than the difference between user utility and the highest price users may accept, while expression (7.23) imposes a similar constraint on the net surplus of the provider.

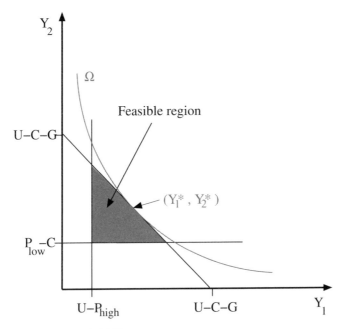

Fig. 7.4 Optimal price point $(Y_1^\star, Y_2^\star)$

Fig. 7.4 plots the curve of the objective function Ω as a function of Y_1 and Y_2. The feasible area is represented by the shaded triangle formed by the linear constraints (7.21)-(7.23). As the value of Ω increases, the curve moves upwards, and vice versa. The maximum value of Ω occurs when the curve intersects the line $Y_1 + Y_2 = U - C - G$ at exactly one point, and the coordinates of this point correspond to the optimal values for Y_1 and Y_2. To obtain the latter values, we rewrite the optimization problem (7.20)-(7.23) in the following Lagrange form:

$$\max_{Y_1, Y_2} \quad \Omega' = Y_1^\beta Y_2^{1-\beta} + \eta(Y_1 + Y_2 - U + C + G) \tag{7.24}$$

where η is the Lagrange multiplier. By taking the partial derivatives of Ω' with respect to Y_1 and Y_2, setting them equal to zero, and using the fact that $Y_1^\star + Y_2^\star = U - C - G$, we obtain:

$$Y_1^\star = \beta(U - C - G) \tag{7.25}$$
$$Y_2^\star = (1 - \beta)(U - C - G) \tag{7.26}$$

From the definition of Y_1 and Y_2, we finally obtain the optimal prices as follows:

$$P_1^\star = (1 - \beta)U + \beta(C + G) \tag{7.27}$$
$$P_2^\star = (1 - \beta)(U - G) + \beta C \tag{7.28}$$

In the case of no transaction costs (i.e., $G = 0$), the price users pay is exactly the amount that the provider receives, hence:

$$P^\star \;=\; P_1^\star \;=\; P_2^\star \;=\; (1-\beta)U + \beta C \tag{7.29}$$

As we can see, the value of β determines the position of the optimal price along the line segment between the utility and cost values.

Let us now consider three special cases with respect to the value of the bargaining parameter β that provide some insight into the optimal solution of the above optimization problem.

Case 1: $\beta = 0$, i.e., the service provider has all the bargaining power; this situation arises whenever the telecommunications market is monopolized by one service provider. In this case we have that $P^\star = U$, hence the service provider enjoys the total social surplus by squeezing out the users' surplus.

Case 2: $\beta = 0.5$, i.e., users and service provider have exactly the same bargaining power. In this case, $P^\star = 0.5(U + C)$, implying that the social welfare is equally shared by the two parties.

Case 3: $\beta = 1$, i.e., the bargaining power resides solely with the users; such a scenario may arise in the telecommunications market when the supply greatly exceeds the aggregate user demand. In this case we have $P^\star = C$, and the provider has to abandon any benefits (provider surplus) from supplying the service.

7.4.2 The Multiple Tier Case

Let us now consider the general case of $p > 1$ tiers of service. We can apply the methodology of the previous subsection to each service tier $z_j, j = 1, \ldots, p$, to obtain the optimal set of prices $P^\star = \{P^\star(z_1), P^\star(z_2), \ldots, P^\star(z_p)\}$. Let us assume for simplicity that the transaction cost G is zero; then, using expression (7.29) we obtain:

$$P^\star(z_j) \;=\; (1-\beta)U(z_j) + \beta C(z_j), \quad j = 1, \ldots, p \tag{7.30}$$

Since both the utility $U(x)$ and the cost $C(x)$ are non-decreasing functions of bandwidth x, we have for $1 \leq j < k \leq p$:

$$P^\star(z_k) \;-\; P^\star(z_k) = (1-\beta)[U(z_k) - U(z_j)] + \beta[C(z_k) - C(z)j)]$$
$$\geq 0. \tag{7.31}$$

In other words, the optimal price is a non-decreasing function of the service tier index, i.e., with the amount of bandwidth offered to the users, consistent with intuition.

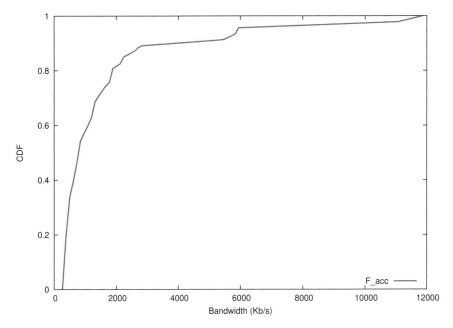

Fig. 7.5 CDF F_{acc} of Internet access speeds (data adapted from [19])

7.5 Performance Evaluation

We illustrate the methodology for sizing and pricing of tiered services that we developed based on concepts from economic theory, by considering the market for
broadband Internet access under either a capacity-based or a usage-sensitive tiered
pricing scheme. Note that the MAX-ES and OPT-P problems require the utility and
cost functions as input, while MAX-ES also requires the PDF of the user demands.
Therefore, we selected these functions to be characteristic of real life scenarios, as
we explain next. The results we present in the remainder of this chapter are representative of the general behavior that we have observed in experiments with a variety
of distribution, utility, and cost functions.

Capacity-based pricing. We have used data collected at the San Diego Network
Access Point (SDNAP) and available at the CAIDA site [19] to obtain the CDF F_{acc}
of Internet access speeds shown in Fig. 7.5. We adapted the raw SDNAP data so that
access speeds are in the range [256 Kbps, 12 Mbps], typical of current broadband
speeds in the United States.

Usage-sensitive pricing. We make the assumption that the monthly amount of traffic generated by users is in the range [5MB, 1TB] and follows the bounded Pareto
distribution (PDF):

$$f(x) = \frac{\alpha k^{\alpha}}{1 - \left(\frac{k}{q}\right)^{\alpha}} x^{-\alpha-1}, \quad 5 = k \leq x \leq q = 10^6, \quad 0 < \alpha < 2 \qquad (7.32)$$

We have selected two values for parameter α, corresponding to two distribution functions:

- PDF $f_{15/85}$ has $\alpha = .00001$ and is such that approximately 15% of users generate about 85% of the total traffic, and
- PDF $f_{5/50}$ with $\alpha = .03$, for which 5% of users generate approximately 50% of the overall traffic.

The latter distribution has characteristics similar to the usage patterns reported recently by one major cable ISP [50, 51].

For all instances of the MAX-ES problem we investigate in this study, we let the utility function be

$$U(x) = \lambda x^{\gamma} \log(x) \qquad (7.33)$$

and the cost function

$$C(x) = \mu x, \qquad (7.34)$$

hence the social surplus function used for solving the MAX-ES problem is given by:

$$S(x) = U(x) - C(x) = \lambda x^{\gamma} \log(x) - \mu x. \qquad (7.35)$$

The utility function (7.33) is an increasing, strictly concave, and continuously differentiable function of service level x, and has also been considered in the context of elastic traffic [105]. The parameters λ and γ can be used to control the slope of $U(x)$. In the following experiments, we use the values $\lambda = 12$, $\gamma = 0.5$, and $\mu = 0.4$ for capacity-based pricing, and $\lambda = 9$, $\gamma = 0.5$ and $\mu = 0.05$ for usage-sensitive pricing, to ensure that the surplus function exhibits similar behavior across the different domains of the corresponding distributions.

7.5.1 Convergence of the Approximate Solution

Let us first consider the impact of the number K of intervals in the PDF approximation (refer to Fig. 5.3) on the convergence of the dynamic programming algorithm we presented in Section 7.3.3. Fig. 7.6 plots the value of the optimal solution $\Phi(K, p)$ in (7.17)-(7.18) as a function of K for three values of $p = 5, 10, 20$, the CDF F_{acc} of Fig. 7.5 and the surplus function defined in expression (7.35). Fig. 7.7 is similar, but shows results for the Pareto CDF $F_{5/50}$.

We make two important observations from these figures. First, for a given number p of service tiers, the solution computed by the dynamic programming algorithm $\Phi(K, p)$ varies widely for small values of K, but converges quickly as K increases. When K is small, the PDF approximation is not accurate, leading to solutions that are far away from the optimal solution to MAX-ES. However, for $K > 100$ the sur-

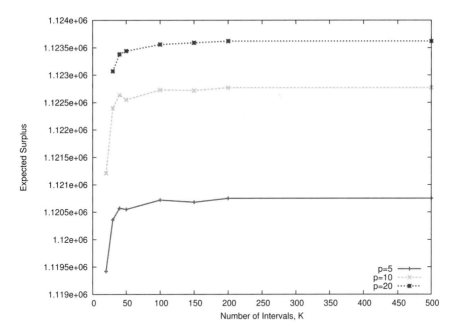

Fig. 7.6 Expected surplus against the number of intervals K, CDF F_{acc}, $p = 5, 10, 20$

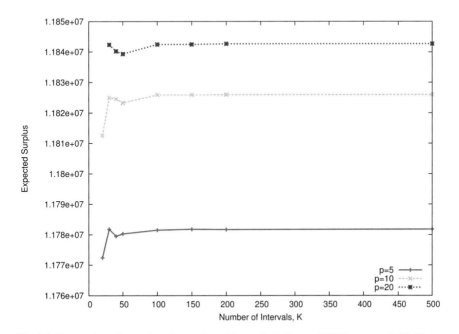

Fig. 7.7 Expected surplus against the number of intervals K, Pareto CDF $F_{5/50}$, $p = 5, 10, 20$

plus curves flatten, indicating that the PDF approximation becomes accurate at this range. We have run experiments with a wide range of instances of MAX-ES beyond the ones we report here, and we have found that in all cases $K = 200$ is sufficient for convergence; hence we have used this value for the experiments we present in the remainder of this section. This result is consistent with our earlier discussion in Chapter 5.3.3 on the convergence of the algorithm for the SDPM1 problem, and confirms that the dynamic programming algorithm provides an accurate and efficient solution to the MAX-ES problem.

We also observe that the expected surplus increases with the number p of service tiers. This behavior is consistent with intuition: a larger number of tiers improves the "resolution" of the final solution and allows the dynamic programming algorithm to better tailor the tiers to the given surplus and distribution functions. A similar behavior is evident in Figs. 5.4-5.10 of Chapter 5 that present results for the SDPM1 problem. Of course, since SDPM1 is a *minimization* problem, the curves in Figs. 5.4-5.10 corresponding to larger p values lie *below* those corresponding to smaller p values. The figures also demonstrate the (expected) effect of diminishing returns, as further increases in the number p of service tiers provide smaller improvements to the expected surplus. This effect was also observed in Figs. 5.4-5.10.

7.5.2 Optimal Sizing of Service Tiers

We now compare the performance of four tiered structures in terms of the expected social surplus they achieve:

1. **Optimal:** this is the structure obtained as the optimal dynamic programming solution to the corresponding Approximate-MAX-ES problem instance.
2. **Optimal-rounded:** this is the tier structure derived by rounding the values of the optimal tiers of the above solution to the nearest multiple of 256 Kbps (for capacity-based pricing) or 10 GB (for usage-sensitive pricing). The motivation for this solution arises from considerations related to marketing the service to customers who do not have intimate knowledge of the manner in which the optimal tier structure is determined. More specifically, a tier of, say, 100 GB, is likely to seem more natural and understandable to users compared to the outcome, say, 98.54 GB, of the dynamic programming algorithm, which could well be considered arbitrary.
3. **Uniform:** in this tier structure, the p service tiers are spread uniformly across the domain $[x_{min}, x_{max}]$, i.e.,

$$z_l = x_{min} + l \frac{x_{max} - x_{min}}{p}, \quad l = 1, \ldots, p.$$

4. **Exponential:** each tier provides a level of service that is twice that of the immediately lower tier:

$$z_{l+1} = 2z_l, \quad l = 1, \ldots, p-1.$$

As a result, the tiers divide the domain $[x_{min}, x_{max}]$ into intervals of exponentially increasing length.

The uniform and exponential tier structures are simple, straightforward solutions that do not involve any optimization and are along the lines of the structures employed by major ISPs[2]. We consider them here as baseline cases and to demonstrate that they perform poorly in terms of maximizing the expected social surplus.

Since the raw expected surplus values are not comparable across different instances of the MAX-ES problem, we introduce the concept of *normalized expected surplus* to illustrate the relative performance of the four algorithms above. For a given problem instance, let V_{max} be the maximum expected surplus value achieved by any of the four tiered structures above over all values of the number p of tiers evaluated in the experiments. If $\hat{V}$ is the expected surplus for a given algorithm-p pair, we define the normalized expected surplus for this pair as:

$$0 \leq \text{Normalized expected surplus} = \frac{\hat{V}}{V_{max}} \leq 1. \qquad (7.36)$$

The normalized expected surplus takes values in $[0,1]$ for all instances of MAX-ES and provides insight into the relative behavior of the four tiered structures. In particular, the closer this quantity is to 1, the better the solution.

The three Figs. 7.8-7.10 plot the normalized expected surplus as a function of the number p of service tiers for the three distribution functions F_{acc}, $F_{5/50}$, and $F_{15/85}$, respectively. Each figure contains four curves, each corresponding to one of the tiered structures for the MAX-ES problem described above. We observe that the curves for the optimal and optimal-rounded solutions almost overlap, and exhibit the best performance by far across all the values of p except very small ones, regardless of the underlying distribution function. In particular, the surplus of the exponential tiering solution decreases rapidly for $p > 2$ to about 30-50% of the optimal expected surplus, depending on the distribution (F_{acc} or Pareto). These results demonstrate that exponential grouping of customers, though favored by ISPs, performs far from optimal from an economic standpoint. In fact, the uniform tiering structure performs better than the exponential one, but it can also be far from the optimal solution for other than very small values of p.

The main conclusion from the results shown in Figs. 7.8-7.10 is that by employing the dynamic programming algorithm (7.17)-(7.18), which has low computational requirements, it is possible to obtain optimal tiering structures that improve the expected surplus over simple uniform or exponential tiering solutions by factors of 2-3. More importantly, the optimization approach we described in this chapter makes it possible to re-optimize the tiering structures over time by adapting the utility, cost, and demand distribution functions as necessary to accommodate evolving user demands and market conditions.

---

[2] Many ADSL providers offer download speeds that follow an exponential tiering structure, e.g., 384 Kbps, 768 Kbps, 1.5 Mbps, 3 Mbps, etc. Similarly for the 5/10/20/40 GB tiers of monthly traffic used in the recent pilot program by a cable ISP [50,51].

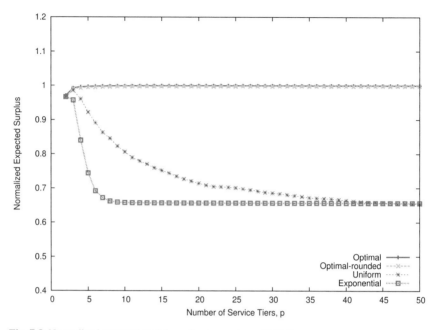

Fig. 7.8 Normalized expected social surplus comparison, CDF F_{acc}

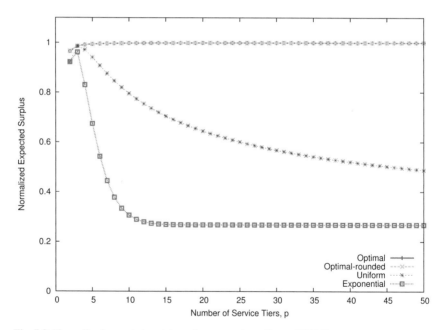

Fig. 7.9 Normalized expected social surplus comparison, Pareto CDF $F_{5/50}$

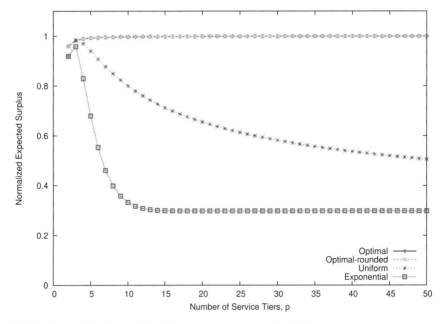

Fig. 7.10 Normalized expected social surplus comparison, Pareto CDF $F_{15/85}$

7.5.3 Optimal Pricing of Service Tiers

Given a set $\{z_1,\ldots,z_p\}$ of p of service tiers, as well as the utility $U(x)$ and cost $C(x)$ functions, we may solve the OPT-P optimization problem we described in Section 7.4 to obtain the optimal prices for each of the service tiers. Fig. 7.11 plots the optimal prices for the $p = 5$ optimal service tiers obtained for CDF F_{acc}. The y axis represents *normalized prices* in that the prices of all tiers are shown relative to the price of the highest tier, such that $P(z_5) = 1$. The x axis represents the range of user requests, i.e., the domain [256 Kbps, 12 Mbps] of CDF F_{acc}. The step jumps in the price curves indicate the values of the optimal service tiers.

As we can see, the optimal tier structure does not resemble either the uniform or exponential structures, confirming our earlier comments on the suboptimality of these solutions. Three price structures are shown, corresponding to the three values of the bargaining power of users $\beta = 0.25, 0.5, 0.75$. As expected, the lower the bargaining power of users, the higher the corresponding price. Also, for a fixed value of β, the price increases with the tier index, consistent with our discussion in Section 7.4.2. Moreover, the price increase from one tier to the next is tied directly to the shape of the utility and cost functions, thus reflecting the perspective of both users and providers.

In the previous subsection we demonstrated that the exponential and uniform tiering structures are far from optimal with respect to the expected social surplus. We now show that these structures are also suboptimal in terms of the revenue collected

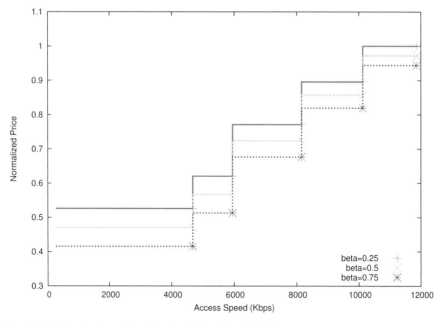

Fig. 7.11 Optimal prices of $p = 5$ optimal price tiers, CDF F_{acc}

by the service provider. Consider a tier set $\{z_1, \ldots, z_p\}$, and let $P(z_j), j = 1, \ldots, p$, be the optimal price structure obtained by applying the methodology of Section 7.4. Then, the expected revenue R collected by the service provider can be calculated as:

$$R(z_1, \ldots, z_p) \;=\; \sum_{j=1}^{p} \left(P(z_j)\left(F(z_j) - F(z_{j-1})\right) \right). \tag{7.37}$$

Figs. 7.12 and 7.13 plot the *normalized expected revenue* against the number p of service tiers for the four tiered structures we described earlier. The normalized expected revenue is defined similarly to the normalized expected surplus in expression (7.36). Note that the highest revenue is obtained when there is only one tier, in which case all users are mapped to the highest possible service (that also incurs the highest price); such a solution is unlikely to be adopted in a market environment, and is included here for illustration purposes only. As the number p of tiers increases, the expected revenue decreases for a while and then stabilizes. The curves for the optimal and optimal-rounded solutions both converge quickly to a value that is around one-half that of the maximum revenue for $p = 1$. However, the exponential and uniform solutions drop much more rapidly, eventually reaching a value that is only one-quarter (for F_{acc}) or one-sixth (for the Pareto distribution) of the maximum revenue. Again, the uniform tiering structure outperforms the exponential one, while the optimal solution achieves an expected revenue that is up to 2-3 times higher than the other two, consistent with the results of the previous section.

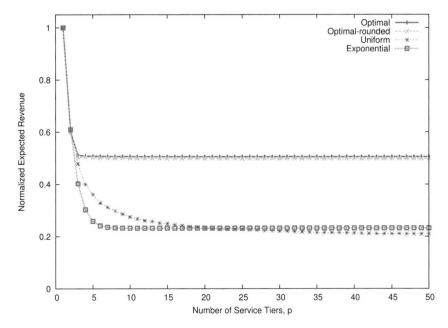

Fig. 7.12 Normalized revenue comparison, CDF F_{acc}

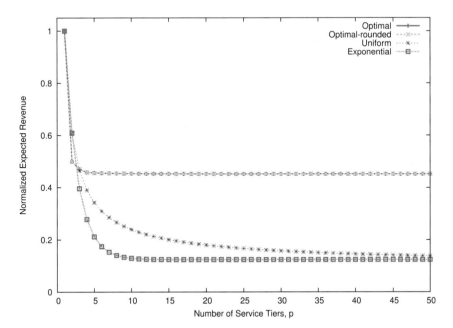

Fig. 7.13 Normalized revenue comparison, Pareto CDF $F_{5/50}$

Since the cost of providing the service is the same regardless of what tiered structure is selected, these results indicate that by adopting simple, suboptimal solutions, the service provider may end up foregoing a substantial fraction of potential revenues. More importantly, these additional revenues are *not* at the expense of users, but rather due to the larger surplus achieved by the optimal solution. In other words, the optimal tiered structure provides substantial more value to both users and providers.

7.5.4 Accounting for the Cost of Service Tiers

Finally, let us investigate how the results are affected when we include the cost of providing each service tier into the optimization. Specifically, Fig. 7.14 plots the normalized expected surplus of the four tiering structures under CDF F_{acc}, against the number p of tiers. These results were obtained using the same utility function as in Fig. 7.8, but the cost function (7.34) was modified to include an incremental per-tier cost α; in other words, the cost function for p service tiers is $C(x) = \mu x + \alpha p$, and we let $\alpha = 5$ here. Results similar to those shown in Fig. 7.14 were obtained for the Pareto distributions and are omitted.

We observe that the relative behavior of the four curves in Fig. 7.14 is similar to that in Fig. 7.8 were it was assumed that there is no cost in offering an additional tier of service. The main difference between the corresponding curves in the two figures is the obvious downwards slope introduced as a result of including a per-tier cost. The curves in Fig. 7.8 eventually level off as the number p of tiers increases, indicating that, beyond a certain point, adding more tiers does not affect the expected surplus. Fig. 7.14 further demonstrates that the tier-related cost starts to dominate as p increases: after reaching a maximum at a certain value of p, all curves start to trend downwards at a slope that eventually becomes equal to $-\alpha$. Clearly, the value of p at which the maximum is reached depends on the parameters of the utility and cost functions; nevertheless, it is also obvious that, overall, the uniform and exponential tiering structures perform poorly compared to the optimal solution.

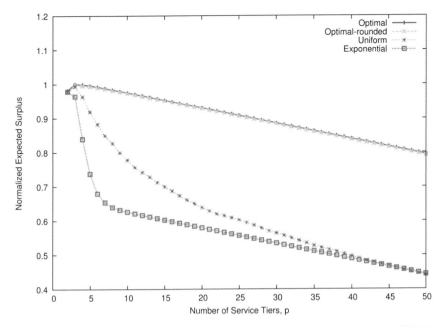

Fig. 7.14 Normalized expected social surplus comparison, taking tier cost into account, CDF F_{acc}

Chapter 8
Service Tiering As A Market Segmentation Strategy

In the previous chapter we considered a market scenario in which all users receive the same value from the service offered by the network operator, or equivalently, all users are characterized by the same utility function $U(x)$. A market in which all users value a service (or product) similarly is said to be *inelastic* [20]. Certain essential goods (e.g., gasoline or milk) and services that everyone needs tend to be inelastic, at least in the short term. Markets for most other products and services tend to be *elastic*, in that the value of the product or service may be perceived quite differently across the population of consumers. Hence, in elastic markets, consumer behavior with respect to pricing may vary widely depending on the underlying utility curve that characterizes the value that the specific consumer places on the product or service in question.

In this chapter we consider broadband Internet access as an elastic service whose value varies across segments of the user population. Specifically, we assume that users are partitioned into classes such that users in a given class are characterized by a distinct utility function. We study the problem of selecting jointly the set of service tiers and their prices so as to maximize the profit (i.e., provider surplus) of the ISP. For the special case of a single tier (i.e., no tiering), we develop an optimal algorithm to determine both the level of service to be offered and its price. We then show that introducing multiple tiers can be an effective market segmentation strategy that can lead to an increase in profits. If the service levels are given and not subject to optimization, we show that the optimal price structure for these tiers to maximize profits may be obtained through an efficient dynamic programming algorithm. We also extend this algorithm to obtain an approximate yet accurate solution to the joint service level-price optimization problem.

8.1 Economic Model of User Diversity

As in Chapter 7, we consider the market for broadband Internet access with one network provider (ISP) and multiple users (consumers). The service provided by the

G.N. Rouskas, *Internet Tiered Services*, DOI: 10.1007/978-0-387-09738-1_8,
© Springer Science + Business Media, LLC 2009

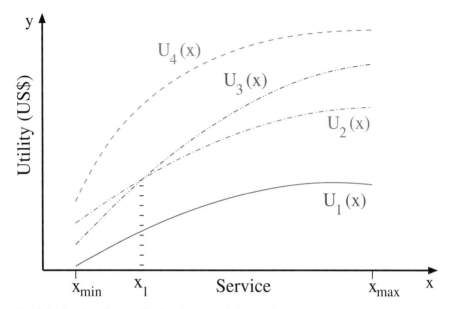

Fig. 8.1 Diversity of user utility functions, $T = 4$ classes of users

ISP is described by a single parameter, namely, access speed x, with x taking values
in the interval $[x_{min}, x_{max}]$, where x_{min} and x_{max} correspond to the lowest and highest
speed, respectively, that the ISP may offer. The cost to the ISP of offering an amount
x of service is given by the cost function $C(x)$. The ISP offers a tiered bandwidth
service with p tiers. As in earlier chapters, we let $Z = \{z_1, \ldots, z_p\}$ denote the set of
distinct service tiers, labeled such that $z_1 < \ldots < z_p$. We also let $P(z_j), j = 1, \ldots, p$,
denote the price the ISP charges users that subscribe to tier z_j. Price is an increasing
function of service x, hence, $i < j$ implies $P(z_i) < P(z_j)$. For notational convenience,
we assume the existence of a "null" service tier z_0 for which $P(z_0) = 0$, and also
that $P(z_{p+1}) = \infty$; this last assumption ensures that no user may receive service at
an amount higher than the highest available tier.

Users of the service belong to one of T classes, $T > 1$. Users within class $t, t =
1, \ldots, T$, are characterized by utility function $U_t(x)$; without loss of generality, we
make the assumption that $U_t(x) \neq U_s(x)$ whenever $t \neq s$. As in the previous chapter,
we express cost, price, and utility in the same units, e.g., US\$. The various utility
curves indicate the users' willingness to pay, and can be determined using traditional
market research instruments such as surveys, or more sophisticated tools such as
conjoint analysis [80]. Fig. 8.1 illustrates the user diversity with respect to utility
curves for $T = 4$ classes. We let $f_t, t = 1, \ldots, T$, denote the fraction of the user
population that is in class t; obviously, $f_1 + \ldots + f_T = 1$.

For the remainder of this chapter, we make the reasonable assumption that the
cost $C(x)$ and utility functions $U_t(x), t = 1, \ldots, T$, are continuous, twice differentiable, and non-decreasing in the interval $[x_{min}, x_{max}]$.

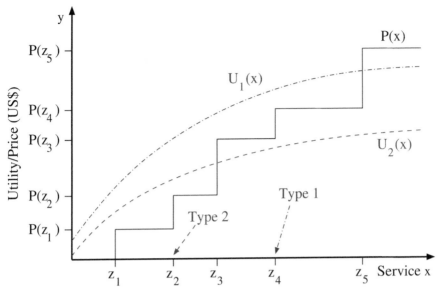

Fig. 8.2 Mapping of $T = 2$ classes of users to $p = 5$ service tiers based on the price structure $P(x)$

Users in the same class t receive the same value from the service, and consequently, they respond identically to the pricing policy of the network provider. Specifically, we make the fundamental assumption that if the price set for a product is *below* the value of this product to a consumer, then the consumer will purchase the product. On the other hand, if the price of the product is *higher* than the consumer's perceived value of the product, then they will not make the purchase. In the tiered service scenario we are considering, the above assumptions imply that users in class t, will subscribe to the highest tier z_j for which the price charged does not exceed its value to the user: More formally, given the utility and price functions, there is an implied mapping $h : \{1,\ldots,T\} \rightarrow Z$ from the set $\{1,\ldots,T\}$ of user classes to the set of service tiers, where $h(t) = z_j$ if and only if:

$$P(z_j) \leq U_t(z_j) < P(z_{j+1}), \quad t = 1,\ldots,T, \quad j = 0,1,\ldots,p. \tag{8.1}$$

We note that the above expression imposes a set of directionality constraints similar to (2.15) for the DPM1 problem, but in the *opposite* direction. In particular, if the price of the lowest tier is higher than the utility of some class of users, then, from (8.1) these users are forced to "subscribe" to the "null" service tier z_0, which implies that they will not use the service.

Fig. 8.2 illustrates the mapping of $T = 2$ classes of users to $p = 5$ service tiers based of the given price structure imposed by the step pricing function $P(x)$. Specifically, users in class 1 and class 2 are mapped to tiers z_4 and z_2, respectively, consistent with expression (8.1).

From the point of view of the ISP, there is a clear tradeoff in setting the price for the service tiers. If the price for some tier is high, the ISP will lose revenue as some customers may decide to subscribe to a lower tier or not use the service at all (if their utility is lower than the price of the lowest tier). On the other hand, if the ISP prices the tiers conservatively, it may attract some low-utility customers, but may also forego a signi cant amount of revenue from customers with high utility who would be willing to pay more for the service. If users do not communicate with each other, then the ISP may set a different level of service and price for each class of users, *independently* of the other classes. This is not a realistic assumption, however, as users do communicate, hence one tiered price structure must be set for all users. Therefore, the objective is to select jointly the p service tiers to be offered and their prices so as to maximize the provider surplus. We will refer to this problem as the *surplus maximization (MAX-S)* problem, and formally de ne it as follows.

Problem 8.1 (MAX-S). Given the provider cost function $C(x)$, an integer T of user classes, the fraction f_t of users in class t and their utility $U_t(x), t = 1, \ldots, T$, and the domain $[x_{min}, x_{max}]$ of the cost and utility functions, nd a set $Z = \{z_1, \ldots, z_p\}$ of p service tiers and their respective prices $P(z_j)$ that maximize the following objective function that represents the provider surplus:

$$S_{pr}(z_1, \ldots, z_p) = \sum_{j=1}^{p} [P(z_j) - C(z_j)] \sum_{h(t)=z_j} f_t \qquad (8.2)$$

under the constraints:

$$h(t) = z_j \text{ iff } P(z_j) \leq U_t(z_j) < P(z_{j+1}), \quad t = 1, \ldots, T, \ j = 0, 1, \ldots, p \quad (8.3)$$

$$0 = z_0 < x_{min} \leq z_1 < z_2 < \ldots < z_p \leq x_{max} \qquad (8.4)$$

The following lemma states that the price of each service tier $z_j, j = 1, \ldots, p$, in the optimal solution to the MAX-S problem may take one of T distinct values.

Lemma 8.1. *Let $Z = \{z_1, \ldots, z_p\}$ be an optimal solution to the MAX-S problem and $P(z_j), j = 1, \ldots, p$, be the price of the corresponding tiers. Then:*

$$\exists t \in \{1, \ldots, T\} : P(z_j) = U_t(z_j), \quad j = 1, \ldots, p. \qquad (8.5)$$

Proof. By contradiction. Assume that the price of tier z_j is such that $U_s(z_j) < P(z_j) < U_t(z_j)$ for some classes $s \neq t$. In other words, class-t users (and, perhaps, users of some class q with $U_q(z_j) > U_t(z_j)$) subscribe to tier z_j, whereas users of class s and any class r with $U_r(z_j) < U_s(z_j)$ do not subscribe to tier z_j. Therefore, the price of the tier can be raised to $P'(z_j) = U_t(z_j)$ without affecting the set of users subscribing to this or any other tier. This increase in price will result in an increase to the provider surplus, contradicting the assumption about optimality of the original price $P(z_j)$. □

Let us now consider the users' point of view. For a given set Z of service tiers and a pricing strategy set by the ISP, the user surplus can be computed as

$$S_{usr}(z_1,\ldots,z_p) = \sum_{t=1}^{T} f_t \left[U_t(z_{h(t)}) - P(z_{h(t)}) \right], \tag{8.6}$$

where $h(t)$ is defined in expression (8.3). The quantity in (8.6) can be viewed as the amount by which users are *richer* by using the service. Indeed, class-t users of some tier z_j value the service at $U_t(z_j)$ but they only paid $P(z_j)$ for it; hence, collectively the users are richer after subscribing to the service than before by an amount equal to the user surplus in (8.6).

8.2 The Single Tier Case

To gain insight into the MAX-S problem, let us first consider the simpler case of $p = 1$, i.e., the provider offers only one level of service z_1. Due to Lemma 8.1, we know that $P(z_1) = U_t(z_1)$ for some t, and our goal is to determine an optimal value for $z_1 \in [x_{min}, x_{max}]$ and a corresponding optimal price. To this end, we distinguish two cases, discussed next.

Case 1. The T utility functions $U_t(x), t = 1,\ldots,T$, and the cost function $C(x)$ do not pairwise intersect anywhere in their domain $[x_{min}, x_{max}]$. Without loss of generality, we make the assumption that $C(x)$ lies below all of the T utility functions in the same interval. If that is not true, we can ignore the utility functions that lie below $C(x)$ and only consider the $T' < T$ functions that lie above $C(x)$. Doing so will not affect optimality, since setting the price below cost will result in a loss for the provider.

Now let us relabel the T utility functions such that:

$$C(x) < U_1(x) < U_2(x) < \ldots < U_T(x), \quad \forall x \in [x_{min}, x_{max}], \tag{8.7}$$

and define

$$F_t = \sum_{s=t}^{T} f_s, \quad t = 1,\ldots,T, \tag{8.8}$$

as the fraction of users falling in the classes with utilities equal to, or higher than, that of class t.

If the provider offers a single tier in the amount of z_1^t and prices it according to the corresponding utility of class-t users, then we can write the provider surplus from (8.2) as:

$$S_{pr}^t(z_1^t) = F_t \left[U_t(z_1^t) - C(z_1^t) \right], \quad t = 1,\ldots,T. \tag{8.9}$$

Therefore, we can find the optimal tier $z_1 \in [x_{min}, x_{max}]$ and its price using these two steps:

1. For each class $t, t = 1, \ldots, T$, determine the value of $z_1^t \in [x_{min}, x_{max}]$ that maximizes the quantity $S_{pr}^t(z_1^t)$ in expression (8.9).
2. Set the tier z_1 to the quantity z_1^q, and its price to $U_s(z_1^q)$, where q is such that $S_{pr}^q(z_1^q)$ is maximum among the T quantities computed in Step 1.

Case 2. Some of the utility and cost functions pairwise intersect in one or more points within their domain $[x_{min}, x_{max}]$. In this case, it is always possible to partition this interval into sub-intervals within which none of the functions intersect. Returning to Fig. 8.1, we can see that the domain of the utility functions can be divided into two sub-intervals, $[x_{min}, x_1]$ and $[x_1, x_{max}]$, within which the functions do not intersect. We also note that the $T + 1$ utility and cost functions may intersect pairwise at only a finite number of points, hence the number K of such sub-intervals is also finite. Therefore, we can obtain the optimal value for z_1 and the corresponding price by following the following steps:

1. Divide the interval $[x_{min}, x_{max}]$ into an appropriate number K of non-overlapping sub-intervals $e_k, k = 1, \ldots, K$, such that no two utility or cost functions intersect within each sub-interval e_k.
2. For each sub-interval e_k, find the optimal value $z_1^q(k)$ and optimal price $U_q(z_1^q(k))$, following the approach we described in Case 1 above.
3. Set the tier z_1 and its price to the corresponding values for the interval e_k with the maximum provider surplus $S_{pr}^q(z_1^q(k))$ among all the intervals in Step 2.

Based on the above discussion, in order to find the optimal solution to the MAX-S problem for $p = 1$ tier, we need to determine the maximum of the provider surplus function in expression (8.9) in any sub-interval $[x_1, x_2]$ of $[x_{min}, x_{max}]$. Let us define function $g(x)$ as:

$$g(x) = U_t(x) - C(x), \quad x \in [x_1, x_2], \tag{8.10}$$

for some class t. Since the utility and cost functions are continuous and twice differentiable throughout their domain, then function $g(x)$ is continuous and twice differentiable in any sub-interval $[x_1, x_2]$. Therefore, we can find its maximum as follows:

1. The second derivative $g''(x) \leq 0$ everywhere in $[x_1, x_2]$. Then, $g(x)$ is concave in the sub-interval, and its maximum can be found by solving the equation $g'(x) = 0$.
2. The second derivative $g''(x) \geq 0$ everywhere in $[x_1, x_2]$. Then, $g(x)$ is convex in the sub-interval, and its maximum values occur at either end-point, x_1 or x_2, of the sub-interval.
3. The second derivative changes sign one or more times in the sub-interval. In this case, we further subdivide $[x_1, x_2]$ into intervals within which the second derivative g'' is either non-negative or non-positive everywhere in the smaller intervals. We obtain the maximum of $g(x)$ within each smaller interval according to either case 1 or case 2 above, from which we can select the overall maximum in $[x_1, x_2]$.

8.3 The Multiple Tier Case: Market Segmentation

Let us now return to the general case of a tiered service with $p > 1$ tiers. Such a service can be viewed as a *market segmentation* strategy [116], whereby the ISP splits the market into several segments with the goal of increasing profitability. A typical example of market segmentation is when providers offer a "premium" service at a high price for the high end of the market, and a "standard" service at a lower price for the rest of the market. An important issue that arises in the market segmentation is process is determining how to segment the market and how to differentiate among the services to be offered to the various segments (e.g., how to differentiate a "premium" from a "standard" service) so as to maximize profitability. A tiered service and price structure obtained as a solution to the MAX-S problem resolves this issue since the tiers and corresponding prices uniquely identify the market segments that optimize the provider surplus (profit).

We also note that market segmentation follows the law of diminishing returns [116] in that, after an initial increase in profits, creating an additional market segment may have a negligible effect in overall profitability. Therefore, the number p of market segments (service tiers) will, in general, be less than the number T of user classes, especially if T is relatively large. In other words, an optimal market segmentation strategy may combine several user classes into a single segment. On the other hand, because of Lemma 8.1, in an optimal solution to the MAX-S problem the price of each tier $z_j, j = 1, \ldots, p$, is equal to the utility $U_t(z_j)$ of some class t. Therefore, the solutions we develop are for the general case $p \leq T$.

For simplicity, in the remainder of this chapter we make the assumption that the T utility curves do not intersect anywhere in the domain $[x_{min}, x_{max}]$ and are labeled such that $U_1(x)$ is the lowest and $U_T(x)$ the highest curve. This is a reasonable assumption, since if some user A values an amount of service x_1 more than a user B, then an amount $x_2 > x_1$ of service is likely to have more value for A than for B. On the other hand, the cost function $C(x)$ may intersect with some of (or all) the utility curves within the interval $[x_{min}, x_{max}]$, but may intersect at most once with a given utility function.

8.3.1 The MAX-S Problem with Fixed Tiers

Let us first consider a restricted version of the MAX-S problem in which the p tiers are provided as input to the problem and are not subject to optimization. This problem variant arises naturally for the simple uniform and exponential tiering structures we discussed in Chapters 4 and 7. Given the minimum and maximum service levels, x_{min} and x_{max}, respectively, and the number p of tiers, the p service levels are completely specified under uniform and exponential tiering, hence only the prices of these levels need to be optimized.

Due to the assumption that utility curves do not intersect and that price is an increasing function of service x, a feasible solution to the MAX-S problem is such

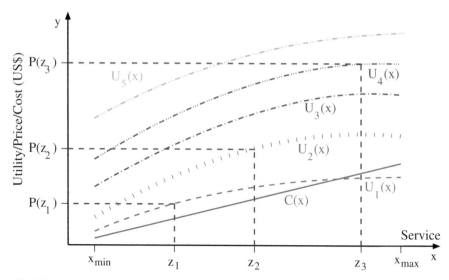

Fig. 8.3 Feasible price structure for an instance of MAX-S with $T = 5$ classes of users and $p = 3$ fixed service tiers

that:

$$P(z_j) = U_s(z_j) \text{ and } P(z_{j+1}) = U_t(z_{j+1}) \Rightarrow s < t, \quad j = 1,\ldots,p-1, \quad (8.11)$$

as illustrated in Fig. 8.3. Therefore, we can obtain an optimal price structure for the MAX-S problem with p fixed tiers using the dynamic programming algorithm described next.

Let $\Lambda(t,l)$ denote the optimal provider surplus when there are t classes of users and l service tiers. For an instance of MAX-S with T classes and p tiers, we can compute $\Lambda(T,p)$ recursively using the following expressions (recall that the quantity F_t is defined in expression (8.8)):

$$\Lambda(t,1) = \max_{1 \le s \le t} \{F_s[U_s(z_1) - C(z_1)]\}, \quad t = 1,\ldots,T \quad (8.12)$$

$$\Lambda(t,l+1) = \max_{s=l,\ldots,t-1} \{\Lambda(s,l) + F_{s+1}[U_{s+1}(z_{l+1}) - C(z_{l+1})]\}$$

$$l = 1,\ldots,p-1, \, t = 2,\ldots,T. \quad (8.13)$$

Expression (8.12) states that, if there is a single service tier fixed at z_1, the price is set to the utility that maximizes the provider surplus (refer also to the definition of provider surplus for a single tier in expression (8.9)). The recursive expression (8.13) is derived from the observation that if the price of the $(l+1)$-th tier is set to the utility of the $(s+1)$-th class, then all users in this class and all classes of higher utility will subscribe to this tier. The second term within the brackets in the right-hand side of (8.13) represents the contribution of this tier to the provider surplus. The first term in brackets in the right-hand side of (8.13) represents the optimal surplus for

s classes of users and l tiers, $r \geq l$. Taking the maximum over all values of s yields the overall maximum.

The running time complexity of the dynamic programming algorithm (8.12)-(8.13) is $O(pT^2)$.

8.3.2 Approximate Solution to the MAX-S Problem

We now turn our attention to the original version of the MAX-S problem whereby both the service level at each tier and its price are subject to optimization. We solve this problem approximately by employing a technique we have used in earlier chapters. Specifically, we divide the interval $[x_{min}, x_{max}]$ into $K > T$ segments of equal length, and impose the additional constraint that the p tiers, $z_1, \ldots, z_p$, may only take values from the set $\{e_k, k = 1, \ldots, K\}$, where $e_k = x_{min} + \frac{k(x_{max}-x_{min})}{K}$ is the right endpoint of the k-th interval. As $K \to \infty$, this discrete version of MAX-S approaches the original version in which $z_j, j = 1, \ldots, p$, are continuous variables.

Let $\Delta(k,t,l)$ denote the optimal solution to this discrete version of MAX-S when there are k points, t classes, and l tiers. We can compute $\Delta(K,T,p)$ recursively as follows:

$$\Delta(k,t,1) = \max_{1 \leq m \leq k} \left\{ \max_{1 \leq s \leq t \leq m} \{F_s[U_s(e_m) - C(e_m)]\} \right\}$$
$$k = 1, \ldots, K, \ t = 1, \ldots, T, \ t < k \tag{8.14}$$

$$\Delta(k,t,l+1) = \max_{m=l,\ldots,k-1} \left\{ \max_{s=l,\ldots,t-1} \{\Delta(m,s,l) \right.$$
$$\left. + \max_{r=m+1,\ldots,k} \{F_{s+1}[U_{s+1}(e_r) - C(e_r)]\} \right\}$$
$$l = 1, \ldots, p-1, \ t = 2, \ldots, T, \ k = 2, \ldots, K, \ t < k \tag{8.15}$$

When there is only one service tier, it is placed at some endpoint e_m and its price is set at the utility of some class s that maximizes the provider surplus, hence we have expression (8.14). In the general case of k points, t classes, and $l+1$ tiers, the optimal value is obtained by (1) considering the best placement and pricing of l tiers in $m < k$ points and $s < t$ classes, given by $\Delta(m,s,l)$, in which case the best placement and price for tier $(l+1)$ is given by the second line of (8.15), and (2) then taking the maximum over all possible values of m and s, yielding the recursive expression (8.15). The running time complexity of this algorithm is $O(pT^2K^3)$. The algorithm can be easily extended to also optimize the number p of tiers, at a complexity of $O(T^3K^3)$.

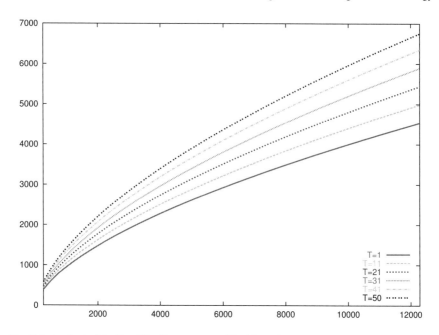

Fig. 8.4 A subset of the $T = 50$ utility curves used in this study

8.4 Performance Evaluation

To evaluate the performance of tiered service as a market segmentation strategy, we consider the market for broadband Internet access. As in Chapter 7 we let the minimum service $x_{min} = 256$ Kbps and the maximum service $x_{max} = 12$ Mbps. For all instances of the MAX-S problem we investigate in this study, we assume the existence of $T = 50$ classes of users characterized by the family of utility curves:

$$U_t(x) = \lambda_t x^\gamma \log(x), \quad t = 1, \ldots, T = 50. \tag{8.16}$$

We use the same cost function as in Chapter 7, i.e.,

$$C(x) = \mu x. \tag{8.17}$$

For the results shown here, we used the following values for parameters μ, γ and λ_t:

$$\mu = 0.3, \quad \gamma = 0.5, \quad \lambda_t = 10 + .1(t-1), \, t = 1, \cdots, T = 50, \tag{8.18}$$

such that the $T = 50$ utility curves do not intersect in their domain [256 Kbps, 12 Mbps] and are labeled from lowest to highest. A subset of these utility curves are shown in Fig. 8.4.

We consider three distributions of users into classes:

- a *uniform* distribution, in which each class contains an equal fraction of the user population:

$$f_t = \frac{1}{T}, \quad t = 1, \dots, T = 50, \tag{8.19}$$

- an *increasing* distribution, such that the fraction of users in a given class increases with utility:

$$f_t = ct, \quad t = 1, \dots, T = 50, \tag{8.20}$$

  where $c = \frac{1}{1275}$ is a constant that ensures that $\sum_{t=1}^{T} f_t = 1$, and
- a *decreasing* distribution, in which the fraction of users in a given class increases with utility:

$$f_t = c(T + 1 - t), \quad t = 1, \dots, T = 50, \tag{8.21}$$

  where $c = \frac{1}{1275}$.

8.4.1 Convergence of the Approximate Solution

Recall that the quality of the approximate solution to the MAX-S problem obtained from the dynamic programming algorithm (8.14)-(8.15) depends on the number K of intervals into which the interval $[x_{min}, x_{max}]$ is partitioned. Fig. 8.5 plots the value of the provider surplus returned by the dynamic programming algorithm as a function of K, for the uniform distribution of users across the $T = 50$ utility classes; three curves are shown, for values of $p = 5, 10, 20$. The results are similar to those shown in Figs. 7.6 and 7.7 in that (1) the dynamic programming algorithm converges quickly with the number K of intervals, and (2) the provider surplus increases with p due to the benefits of market segmentation. These results and similar ones for the increasing and decreasing distribution of users into classes indicate that $K = 100$ is sufficient for convergence, hence we use this value in all the experiments we present next.

8.4.2 Tier Structure Comparison

We again compare the performance of the four tiered structures we investigated in Chapter 7:

1. **Optimal:** the tiered structure obtained from the dynamic programming algorithm (8.14)-(8.15).
2. **Optimal-rounded:** the tiered structure derived from rounding the values of the optimal tiers to the nearest multiple of 256 Kbps.
3. **Uniform:** the p tiers are spread uniformly across the domain [256 Kbps, 12 Mbps].
4. **Exponential:** the p tiers divide the domain [256 Kbps, 12 Mbps] into intervals that double in length (from left to right).

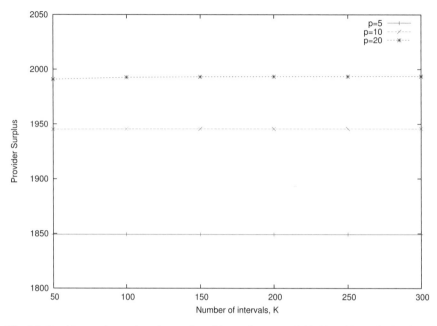

Fig. 8.5 Provider surplus against the number of intervals K, $p = 5, 10, 20$, uniform distribution of users into utility classes

We note that with either uniform or exponential tiering, the $p > 1$ service tiers are completely defined; hence, their prices were optimized using the approach and dynamic programming algorithm we described for fixed tiers in Chapter 8.3.1. For the optimal structure, on the other hand, we obtained both the $p > 1$ service levels and their prices using the dynamic programming algorithm in Chapter 8.3.2. However, for $p = 1$, we obtained the optimal service level and its price following the algorithm in Chapter 8.2, and we use this value for the curves of *all four* tiering structures.

The three Figs. 8.6-8.8 plot the provider surplus against the number p of tiers and correspond to the decreasing, uniform, and increasing distribution of users into classes, respectively. The three figures show four curves, each corresponding to one of the four tiering structures. As we can see, the tiering structure (referred here as "optimal") obtained from the approximate solution to the MAX-S problem and the corresponding optimal-rounded structure outperform the uniform and exponential tiering structures across the range of values for p and across the user distributions into classes. Therefore, network providers would benefit by applying the dynamic programming solutions to determine the tiered structures to offer. Furthermore, although the uniform and exponential tiering structures uniquely define the various tiers to be offered, the prices for these tiers are determined by the dynamic programming algorithm (8.12)-(8.13) so as to optimize the provider surplus for the given tiers. Any other *ad hoc* pricing scheme would result in a lower surplus, hence

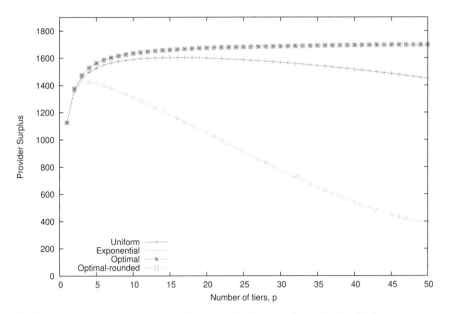

Fig. 8.6 Provider surplus comparison, decreasing distribution of users in $T = 50$ classes

even for these structures the providers would benefit from the tier pricing method-ology presented earlier in this chapter.

There are several more important observations we make from these three fig-ures. First, it is clear that the curves for the optimal and optimal-rounded structures increase rapidly with p initially, but flatten out once the number of tiers increases beyond $p = 10 - 15$; the latter behavior implies that additional tiers provide dimin-ishing returns beyond a point. This result is consistent with economic theory which predicts that there is a limit to the benefits that can be achieved by segmenting a market; it also provides further confirmation to the thesis of this book that a rela-tively small number of service tiers is sufficient to capture most of the benefits of tiering.

We also observe that the exponential tiering structure performs poorly overall, and that its curves reach a well-defined maximum: the surplus achievable under such structure peaks at a small value of p and starts to decline rapidly thereafter. This behavior can be explained by noting that most of the tiers in an exponential structure are grouped together at the leftmost part of the service domain $[x_{min}, x_{max}] = [256$ Kbps, 12 Mbps], and the few tiers that cover the remaining of the interval do not provide fine enough granularity to capture the benefits of market segmentation. The curves corresponding to the uniform tiering structure are below those for the optimal and optimal-rounded structures, but higher than those for exponential tiering. These results indicate that tiering structures with equally spaced tiers would be better for the service provider than exponential ones. Furthermore, we can see that the uniform

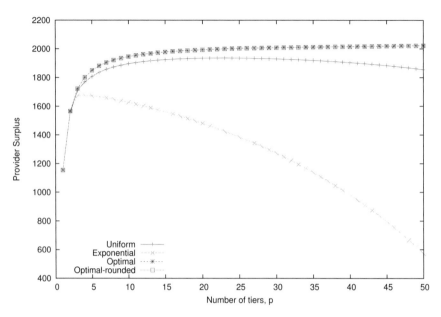

Fig. 8.7 Provider surplus comparison, uniform distribution of users in $T = 50$ classes

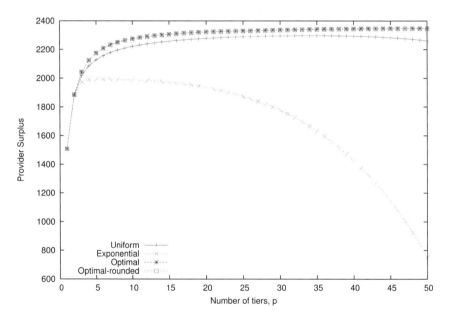

Fig. 8.8 Provider surplus comparison, increasing distribution of users in $T = 50$ classes

tiering curves also reach a maximum at a certain value of p that depends on the user distribution, and start to decline as p increases further. This behavior demonstrates that simply adding more tiers but placing them into specific points in the domain of the service is not an effective market segmentation strategy; hence, to achieve the maximum benefits of market segmentation the service provider must optimize both the size and price of each tier.

Finally, we note that the overall provider surplus increases as we move from the decreasing distribution of users into classes (Fig. 8.6) to the uniform distribution (Fig. 8.7) to the increasing distribution (Fig. 8.8). This is expected, since the fraction of users characterized by high utility functions (i.e., willing to pay higher prices) is lowest for the decreasing distribution and highest for the increasing distribution. As a result, the surplus that the provider is able to extract through market segmentation is higher in the latter case.

Chapter 9
Tiered Service Bundling Under Budget Constraints

The term *product/service bundle* refers to combining several products or services together and selling them as a single package. In Chapter 6, we used concepts from location theory to determine optimal tiering structures for service bundles by formulating the problem as a directional p-median problem in multiple dimensions. Product bundling is also widely used as a marketing strategy [17]. Bundles are often priced at a discount to the total price that their constituent products or services would fetch if they were sold separately. Bundling can be beneficial to both consumers and sellers. The former, in addition to the lower overall price, may appreciate the lower transaction costs and simplified decision process compared with shopping for individual products or services, and may experience a better overall performance due to complementarities among the bundle components. For sellers, bundling has the potential to reduce production and transaction costs, reduce customer churn, and increase revenue and profitability by enabling them to extract a larger part of consumer (user) surplus. In particular, bundling is most successful as a marketing strategy whenever the marginal costs of bundling are low, customer acquisition costs are high, and there are economies of scale in production and distribution of the bundled products. Consequently, bundling is common in industries that share these characteristics, including the telecommunications and cable TV industry, the software business, and the fast food industry, among others.

In this chapter we revisit the problem of determining optimal tiering structures for service bundles using tools from economics and utility theory. Specifically, we assume that the utility of a service bundle to the users is described by a Cobb-Douglas function whose indifference curves impose a preference ranking of service bundles. We also make the assumption that users are limited by budget constraints and that their budgets follow a general distribution. We consider the problem of selecting jointly the tiers and their prices so as to maximize the expected provider surplus under the user budget constraints. We provide approximate dynamic programming algorithms both for the case of predetermined tiers (i.e., when only price is subject to optimization) and the general version of the problem.

G.N. Rouskas, *Internet Tiered Services*, DOI: 10.1007/978-0-387-09738-1_9,
© Springer Science + Business Media, LLC 2009

9.1 Economic Model of Service Bundling

We consider the market for broadband Internet communication services with one ISP and a large number of users. The ISP offers two services. One service, characterized by parameter x (e.g., access speed), may be offered at levels between a minimum x_{min} and a maximum x_{max}. The second service, say, web hosting, is also characterized by a single parameter y (e.g., corresponding to monthly amount of traffic handled), with y also taking values between a minimum y_{min} and a maximum y_{max} level. The ISP bundles the two services into a package, and offers a tiered structure with p tiers for the combined service. We let $Z = \{(z_1,t_1),\ldots,(z_p,t_p)\}$ denote the set of p distinct service tiers, where the j-th tier $(z_j,t_j), t = 1,\ldots,p$, corresponds to an amount z_j for service x and an amount t_j for service y.

We let $C(x,y)$ denote the cost to the ISP of offering a service bundle (x,y) of the two services. We also let $P(z_j,t_j), j = 1,\ldots,p$, denote the price that the ISP charges subscribers to tier (z_j,t_j). Without loss of generality, we assume that tiers are labeled such that

$$P(z_{j-1},t_{j-1}) \quad < \quad P(z_j,t_j), \quad j = 2,\ldots,p. \tag{9.1}$$

For mathematical convenience, we also define the "null" service tier ($z_0 = 0, t_0 = 0$) with price $P(z_0,t_0) = 0$, as well as a fictitious $(p+1)$-th service tier such that $P(z_{p+1},t_{p+1}) = \infty$.

The value that users receive from a bundle (x,y) of the two services is described by the utility function $U(x,y)$. In essence, the utility function imposes a pairwise ranking of bundles by order of preference, where *preference* is a transitive relation. More precisely, if $U(x,y) > U(x',y')$, then bundle (x,y) is said to be strictly preferred to bundle (x',y'). On the other hand, if $U(x,y) = U(x',y')$, the two bundles are equally preferred, and the consumer is said to be *indifferent* between the two bundles. In particular, a curve

$$U(x,y) \quad = \quad u \tag{9.2}$$

is referred to as an *indifference curve* since the user has no preference for one bundle over another among the bundles represented by points along this curve. In other words, each point on an indifference curve provides the same level of utility (value, or satisfaction) to the user. Indifference curves are typically used to represent demand patterns for product or service bundles observed over a population of consumers.

Fig. 9.1 shows a set of indifference curves, each associated with a different utility level. In this figure, utility is measured along the z (vertical) axis, and the indifference curves are simply the projections of the function $U(x,y) = u$, for various values of constant u, on the xy plane. In Fig. 9.1, users would rather be on curve I_7 rather than I_6; they would also rather be on curve I_6 rather than on I_5, and so on, but they do not care where they are on a given indifference curve. Conceptually, indifference curves are similar to topographical maps, in that each point along a given curve is at the same "altitude" above the floor.

The characteristics of the curves in Fig. 9.1 are typical of indifference curves in general. Specifically, indifference curves are defined only on the positive quadrant

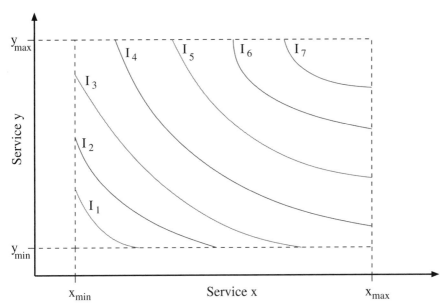

Fig. 9.1 Indifference curves, $I_1, \ldots, I_7$, such that $U(x,y) = $ constant along each curve (utility is measured on the vertical z axis)

of the xy plane, and they are negatively sloped and convex; in other words, as the quantity of one service or good x (respectively, y) that is consumed increases, it must be offset by a decrease in the quantity consumed of the other good y (respectively, x), so as to keep utility (satisfaction) constant.

The Cobb-Douglas family of functions [4] generate indifference curves with the characteristics shown in Fig. 9.1 and are widely used as utility functions in this context[1]. This parameterized family of functions is defined as:

$$U(x,y) \;=\; x^{\alpha} y^{1-\alpha}, \quad 0 \le \alpha \le 1, \tag{9.3}$$

where α is a parameter whose value is used to specify a certain function within the family. Then, the indifference curve for a constant level u of utility is given by:

$$y \;=\; u^{\frac{1}{1-\alpha}} x^{\frac{-\alpha}{1-\alpha}} \tag{9.4}$$

We will use the Cobb-Douglas utility in (9.3) as the utility function in this chapter.

We make the assumption that each user has a budget B, where B is a random variable defined in the interval $[B_{min}, B_{max}]$. We let $f(B)$ and $F(B)$ denote the PDF and CDF, respectively, of random variable B. We make the assumption that a consumer

---

[1] Cobb-Douglas functions are also used as *production functions*, in which case U represents output, x represents capital, and y represents labor. The Cobb-Douglas utility is a special case (with $\rho = 0$) of the more general family of *constant elasticity of substitution* functions defined as $U(x,y) = (\alpha x^{\rho} + (1-\alpha)y^{\rho})^{1/\rho}$, with $\rho \le 1$.

will make a purchase if and only if the price of the product is no greater than the consumer's budget. More specifically, given a set Z of p tiers and a price structure consistent with (9.1), a user will subscribe to the tier (z_j, t_j) with the highest index j whose price $P(z_j, t_j)$ does not exceed the user's budget B.

We are interested in selecting a set of service tiers for the bundled services, and determining their prices, so as to maximize the expected provider surplus. We call this the *maximization of expected provider surplus in two dimensions (MAX-ES-2D)* problem, defined formally as:

Problem 9.1 (MAX-ES-2D). Given the cost and utility functions $C(x,y)$ and $U(x,y)$, respectively, defined in the domain $[x_{min}, x_{max}] \times [y_{min}, y_{max}]$, and the CDF $F(B)$ of user budgets, find a set $Z = \{(z_1, t_1), \dots, (z_p, t_p)\}$ of p service tiers and their respective prices $P(z_j, t_j)$ that maximizes the following objective function representing the expected provider surplus:

$$\bar{Q}_{pr}(Z) = \sum_{j=1}^{p} \left((P(z_j, t_j) - C(z_j, t_j)) \left(F(P(z_{j+1}, t_{j+1})) - F(P(z_j, t_j)) \right) \right) \quad (9.5)$$

under the constraints:

$$P(z_1, t_1) < P(z_2, t_2) < \ \dots \ < P(z_p, t_p) \quad (9.6)$$

$$P(z_j, t_j) \leq U(z_j, t_j), \quad j = 1, \dots, p \quad (9.7)$$

$$x_{min} \leq z_j \leq x_{max}, \quad y_{min} \leq t_j \leq y_{max}, \quad j = 1, \dots, p \quad (9.8)$$

Note that the terms $F(P(z_{j+1}, t_{j+1})) - F(P(z_j, t_j))$, $j = 1, \dots, p$, in the right-hand side of (9.5) represent the fraction of users whose budgets fall in the intervals $[P(z_j, t_j), P(z_{j+1}, t_{j+1}))$, hence they will subscribe to tier j (recall also that we have defined $P(z_{p+1}, t_{p+1}) = \infty$, and that $F(P(z_{p+1}, t_{p+1})) = 1$). Also, constraint (9.7) states that the price of a service tier has to be no greater than the utility (value) of this tier to users, since otherwise users will not subscribe even if their budget allows them to do so.

We have the following result.

Lemma 9.1. *Let* $Z = \{(z_1, t_1), \dots, (z_p, t_p)\}$ *be an optimal solution to the MAX-ES-2D problem. Let* $u_j = U(z_j, t_j)$, $j = 1, \dots, p$. *Then, for all* j, *tier* (z_j, t_j) *is the point on the indifference curve* $U(x, y) = u_j$ *that minimizes the cost* $C(x, y)$.

Proof. By contradiction. Assume that in the optimal solution the j-th tier, $1 \leq j \leq p$, is such that $C(z_j, t_j)$ is not the minimum cost point on the indifference curve $U(x, y) = u_j$. Let (z_j', t_j') be such a minimum cost point, and let Z' be the solution derived from Z with (z_j, t_j) replaced by (z_j', t_j'). Since the utility and price of the

j-th tier is not affected by this change, from (9.5) it is clear that $Q_{pr}(Z') > Q_{pr}(Z)$, contradicting the assumption that Z is an optimal solution. $\square$

9.2 Approximate Solution to the MAX-ES-2D Problem

9.2.1 The Fixed Tier Case

Consider first a special variant of the MAX-ES-2D problem in which the p service tiers are predetermined and part of the input, and not subject to optimization, as in the case of uniform or exponential tiering structures. The cost of each tier is completely determined in this case, and for simplicity we let $C_j = C(z_j, t_j), j = 1, \ldots, p$. Observe also that, if we fix the price of the p-th tier, say at π, the contribution of this tier to the provider surplus is completely determined and depends on the fraction of customers with budgets higher than π. Then, the optimization problem reduces to determining optimal prices for the remaining $p - 1$ tiers that are smaller π.

Recall that tier prices may be chosen from the interval $(0, B_{max}]$, where B_{max} is the maximum value for the random variable B representing the budget. Let us divide this interval into $K > p$ equal-length sub-intervals, such that the right endpoint B_k of the k-th sub-interval is $B_k = \frac{kB_{max}}{K}, k = 1, \ldots, K$. Let $\Upsilon(k, l, w)$ denote the optimal value of (9.5) when there are k sub-intervals, l tiers and the price of the l-th tier is set at $B_w, w \le k$. Then, we may write the following recursion:

$$\Upsilon(k, 1, w) = (B_w - C_1)(F(B_k) - F(B_w)), \quad k = 1, \ldots, K, \ w = 1, \ldots, k \quad (9.9)$$

$$\Upsilon(k, l+1, w) = \max_{q=l,\ldots,w-1} \left\{ (B_w - C_{l+1})(F(B_k) - F(B_w)) + \max_{v=l,\ldots,q} \{\Upsilon(q, l, v)\} \right\}$$
$$l = 1, \ldots, p-1, \ k = 2, \ldots, K, \ w = 1, \ldots, k. \quad (9.10)$$

Expression (9.9) can be explained by noting that when there are k sub-intervals and only one tier with a price set to B_w, the customers who subscribe to the service at this price are those with budgets equal to or greater than B_k, or a fraction $(F(B_k) - F(B_w))$ of the total user population. For each subscriber, the provider has a profit of $B_w - C_1$, hence the expected surplus is given by (9.9). Expression (9.10) can be similarly explained. Once $\Upsilon(k, l, w)$ has been computed for all values of k, l, and w, the overall optimal for p tiers and K intervals can be determined as:

$$\max_w \ \Upsilon(K, p, w) \quad (9.11)$$

The overall running time complexity of this dynamic programming algorithm is $O(pK^2)$.

9.2.2 Cost Minimization on an Indifference Curve

Before we tackle the general version of the MAX-ES-2D problem, we note that, because of Lemma 9.1, each tier in an optimal solution is the point on an indifference curve with the minimum cost among all points on this curve. Therefore, let us consider the optimization problem of the form:

$$\text{Minimize } C(x,y) \tag{9.12}$$

subject to the constraints:

$$U(x,y) = u. \tag{9.13}$$

Depending on the form of the cost and utility functions, this problem may be solved exactly or approximately using standard optimization techniques. Here we will only consider cost functions $C(x,y)$ that are linear functions of x and y:

$$C(x,y) = c_1 x + c_2 y. \tag{9.14}$$

Then, the Lagrange multiplier form of the objective function is:

$$\Phi = C(x,y) + \lambda(U(x,y) - R). \tag{9.15}$$

Assuming that the utility function belongs to the family of the Cobb-Douglas functions in (9.3), we may solve the equations:

$$\frac{\partial \Phi}{\partial x} = c_1 + \lambda \alpha x^{\alpha-1} y^{1-\alpha} = 0 \tag{9.16}$$

$$\frac{\partial \Phi}{\partial y} = c_2 + \lambda(1-\alpha) x^{\alpha} y^{-\alpha} = 0 \tag{9.17}$$

$$x^{\alpha} y^{1-\alpha} = u \tag{9.18}$$

to obtain the values of x and y that minimize the provider's cost along the indifference curve with utility u as:

$$x^{\star} = u \left(\frac{c_1(1-\alpha)}{c_2 \alpha} \right)^{\alpha-1}, \quad y^{\star} = u \left(\frac{c_1(1-\alpha)}{c_2 \alpha} \right)^{\alpha} \tag{9.19}$$

9.2.3 Joint Optimization of Service Tiers and Prices

The most general version of the MAX-ES-2D problem involves the joint selection of service tiers and their respective prices so as to maximize provider surplus, subject to the constraints (9.6)-(9.8). The utility function $U(x,y)$ provides a relative ranking of service bundles (x,y) in terms of user preference, and the utility of any service tier will lie in the interval $[0, U_{max}]$, where $U_{max} = U(x_{max}, y_{max})$. Therefore, the problem can be logically decomposed into three subproblems:

1. find the indifference curve I_j (i.e., utility value $u_j \in [0, U_{max}]$) on which each optimal service tier $(z_j, t_j), j = 1, \ldots, p$, lies;
2. set tier (z_j, t_j) to the point in indifference curve $I_j, j = 1, \ldots, p$, that minimizes the provider cost $C(z_j, t_j)$; and
3. determine the optimal price for each service tier.

We emphasize that this is a logical decomposition that allows us to reason about the problem; in an optimal solution, the three subproblems would be solved jointly and simultaneously.

Given this decomposition, one possible approach to solving MAX-ES-2D approximately would be to divide the interval $[0, U_{max}]$ into M sub-intervals of equal length, and impose the additional constraint that the p tiers be on the indifference curves defined by: $U(x, y) = u_m$, where $u_m = \frac{m U_{max}}{M}, m = 1, \ldots, M$, is the utility corresponding to the right endpoints of the m-th such sub-interval. Consequently, we have to select p of these M endpoints as the service tiers. Let c_m be minimum cost over all points along the indifference curve $U(x, y) = u_m$, and (z_m, t_m) be a point on this curve that achieves this minimum. Then, if the p tiers are given fixed values from the set $\{(z_m, t_m), m = 1, \ldots, M\}$, the optimal prices of these tiers can be obtained using the dynamic programming algorithm we developed in Chapter 9.2.1 for the fixed-tier case. Finally, we can obtain the overall optimal solution under the above additional constraint by evaluating all possible subsets of $\{(z_m, t_m), m = 1, \ldots, M\}$ of cardinality p, and selecting the one, and corresponding pricing structure, with the highest value for the expected provider surplus. Note that the number of subsets that need to be evaluated is equal to $\binom{M}{p}$, and that the number of intervals M has to be large for the approximation to have sufficient accuracy so as to ensure that the obtained solution is of sufficiently high quality. Hence, the computational requirements of obtaining high-quality solutions can be prohibitive.

We now extend the dynamic programming algorithm described in (9.9)-(9.10) to optimally select both the service tiers and their prices. Specifically, we partition the *budget interval* $(0, B_{max}]$, from which tier prices may be chosen, into $K > p$ sub-intervals of equal length, with B_k representing the right endpoint of the k-th sub-interval, $k = 1, \ldots, K$ (recall that B_{max} is the maximum value of the random variable B that represent the users' budgets). We also partition the *utility interval* $[0, U_{max}]$, from which the indifference curves (on which tiers lie) may be chosen, into $M > p$ sub-intervals of equal length, with $u_m, m = 1, \ldots, M$, representing the right endpoint of the m-th sub-interval. Let $\Pi(k, m, l, w)$ denote the optimal value of the expected provider surplus in (9.5) when:

- the number of sub-intervals of the budget interval is k;
- the number of sub-intervals of the utility interval is m;
- the number of tiers is l; and
- the price of the l-th tier is set at $B_w, w \le k$.

We can now write the following recursion. Note that the boundary conditions in (9.20) are a generalization of the corresponding expression (9.9): the optimal value is just the maximum over all possible indifference curves $U(x,j) = u_s$ where the single tier may lie. At each indifference curve, we use the fixed-tier expression (9.9), with the corresponding minimum cost c_s. The recursive expression (9.21) is a similar generalization of (9.10).

$$\Pi(k,m,1,w) = \max_{1 \le s \le m} \{(B_w - c_s)(F(B_k) - F(B_w))\}$$

$$k = 1, \ldots, K, \ m = 1, \ldots, M, \ w = 1, \ldots, k \qquad (9.20)$$

$$\Pi(k,m,l+1,w) = \max_{s=l,\ldots,m-1} \left\{ \max_{q=l,\ldots,w-1} \{(B_w - c_s)(F(B_k) - F(B_w))\} \right.$$

$$\left. + \max_{v=l,\ldots,q} \{\Pi(q,s,l,v)\} \right\}$$

$$l = 1, \ldots, p-1, \quad k = 2, \ldots, K, \ m = 1, \ldots, M, \ w = 1, \ldots, k. \qquad (9.21)$$

Finally, the optimal value for p tiers, and K and M sub-intervals, respectively for the budget and utility intervals, is given as:

$$\max_w \Pi(K,M,p,w) \qquad (9.22)$$

and can be obtained in time $O(pMK^2)$.

9.3 Performance Evaluation

In order to evaluate tiering structures for service bundles, we consider an ISP offering a bundle of two services, namely, access speed x and web hosting traffic handled y. The domain of service x is [256 Kbps, 12 Mbps], while the domain of service y is [100 MB, 1 TB]. We consider the following tiering structures in our study:

1. **Optimal:** the set of tiers $Z = \{(z_1,t_1),\ldots,(z_p,t_p)\}$ obtained as a solution to the dynamic programming algorithm (9.20)-(9.22), where $z_i \in$ [256 Kbps, 12 Mbps] and $t_i \in$ [100 MB, 1 TB].
2. **Optimal-rounded:** the set of tiers obtained after rounding the values of each tier $(z_i,t_i) \in Z$ such that z_i is rounded to the nearest multiple of 256 Kbps and t_i is rounded to the nearest multiple of 100 MB.
3. **Uniform-uniform:** the tier structure constructed by (1) obtaining a uniform tiering structure $\{z_1,\ldots,z_p\}$ for service x by spreading the p tiers across the domain [256 Kbps, 12 Mbps], (2) obtaining a uniform structure $\{t_1,\ldots,t_p\}$ for service y by spreading the p tiers across the domain [100 MB, 1 TB], and (3) pairing the tiers of same index in the two sets to form the tiers $\{(z_1,t_1),\ldots,(z_p,t_p)\}$ for the bundle.
4. **Exponential-exponential:** this tier structure is obtained in a similar manner as uniform-uniform, except that the p single-service tiers divide their respective

domain into exponential intervals (i.e., intervals that double in length, from left to right).

5. **Uniform-exponential:** the tier structure in which p uniform (respectively, exponential) tiers are obtained for service x (respectively, service y), which are then paired to obtain the p tiers for the service bundle.

6. **Exponential-uniform:** the structure in which the tiers for service x are exponential and those of service y are uniform.

For the last four tiering solutions, the $p > 1$ service tiers are fixed. Therefore, we only optimized the price of each tier following the approach we described in Section 9.2.1 and the dynamic programming algorithm (9.9)-(9.11).

In order to study the effect of the distribution of user budgets, we consider three distinct distributions in the domain $[B_{min} = 10, B_{max} = 1000]$:

- a *uniform* distribution with PDF $f(B) = \frac{1}{B_{max}-B_{min}} = \frac{1}{990}$ and mean 495,
- an *increasing* distribution, $f(B) = \frac{2B}{(B_{max}-B_{min})^2} - \frac{2B_{min}}{(B_{max}-B_{min})^2}$, with mean 650, in which the mass of the distribution is concentrated at higher budget values (more affluent population), and
- a *decreasing* distribution, $f(B) = -\frac{2B}{(B_{max}-B_{min})^2} + \frac{2B_{max}}{(B_{max}-B_{min})^2}$, with mean 345, in which the mass of the distribution is concentrated at lower budget values (less affluent population).

We use the Cobb-Douglas utility function in expression (9.3) with parameter $\alpha = 0.6$, and a linear cost function as in expression (9.14), with $c_1 = 0.1$ and $c_2 = 0.01$; these values for the constants c_1 and c_2 were selected so that neither term of the cost function dominates across the domains of services x and y.

Figs. 9.2-9.4 plot the expected provider surplus for the decreasing, uniform, and increasing, respectively, distribution of user budgets. Each figure shows six curves, corresponding to the six tiered structures above. A first observation is that, for a given tiered structure and a given number of tiers, the expected provider surplus depends directly on the distribution of user budgets. Specifically, the provider surplus increases from Fig. 9.2 (decreasing distribution) to Fig. 9.3 (uniform distribution) to Fig. 9.4 (increasing distribution). This result is directly due to the fact that the average user budget is lowest under the decreasing distribution and highest under the increasing distribution.

We also note that the exponential-exponential structure performs significantly better than the uniform-uniform structure. On the other hand, the curves of the two mixed structures, exponential-uniform and uniform-exponential, lie between the other two, with the former slightly outperforming the latter. This relative behavior is consistent across the three budget distributions, and is in contrast to our observations in the last two chapters where the exponential structure was the worst performed by far. This result can be explained by the fact that, in this problem formulation, the surplus for a given tier is defined as the difference between user budget and cost, while previously it was defined as the difference between utility and cost. With the exponential structure, many tiers are concentrated at small values of the services x and y, where cost is quite low, but the dynamic programming algorithm is

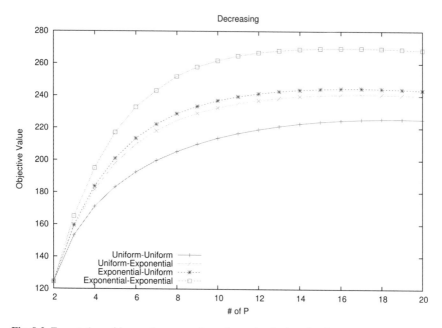

Fig. 9.2 Expected provider surplus comparison, decreasing budget distribution

able to assign these tiers to larger budget values and thus increase the user surplus; this was not possible in the problems we formulated in the previous two chapters, as the utility functions were fixed.

Finally, we observe that the optimal and optimal rounded structures outperform the other four fixed-tier structures, as expected and a strong indication that the optimization methodology we developed in this chapter represents a valuable tool for service providers.

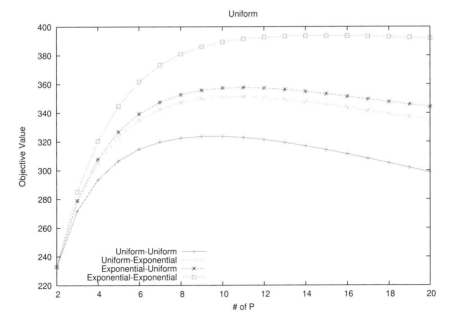

Fig. 9.3 Expected provider surplus comparison, uniform budget distribution

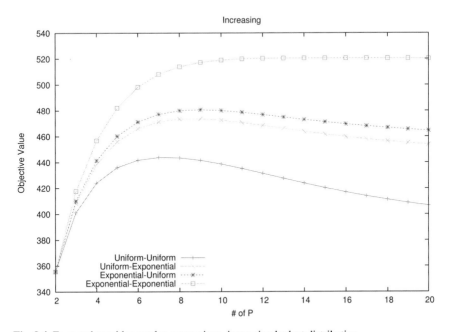

Fig. 9.4 Expected provider surplus comparison, increasing budget distribution

Part III
Quality of Service (QoS)

Chapter 10
Packet Scheduling

In packet-switched networks, packets from various users (flows) have to share the network resources, including buffer space at the routers and link bandwidth. Whenever resources are shared, contention arises among users seeking service. Consequently, shared resources employ a *scheduling discipline* to resolve contention by determining the order in which users receive service. In particular, the scheduling algorithm is a central component of the quality of service (QoS) architecture of packet switched networks. In this chapter, we discuss the requirements for link scheduling disciplines, we review and classify a number of packet schedulers, and we describe their properties and the trade-offs involved. This discussion sets the stage for the introduction in the next chapter of a new, scalable packet scheduler that is based on the concept of tiered service.

10.1 Scheduling Objectives and Requirements

As the Internet has developed into a ubiquitous global communication medium, it is used to carry a constantly evolving mix of applications that is becoming richer as innovation and improvements in technology spawn new services and uses of the network. Nevertheless, network applications can broadly classified into two fundamental classes: *best-effort* and *guaranteed-service*. The two types of applications differ in terms of their sensitivity to delay and availability of bandwidth, as well as in terms of the level of service quality they expect from the network, as we discuss next.

Most of the applications that were originally developed for the Internet (including email, file transfer, and web browsing) and continue to be quite common today, have *elastic* requirements from the network. In other words, such applications are able to adapt to the bandwidth, delay, or loss performance they experience. Elastic applications do not require any explicit guarantees and work correctly with a best-effort service under which the network only makes a promise to attempt to deliver their packets.

G.N. Rouskas, *Internet Tiered Services*, DOI: 10.1007/978-0-387-09738-1_10,
© Springer Science + Business Media, LLC 2009

On the other hand, real-time and interactive applications (including audio and video streaming, multimedia conferencing, etc) do require performance bounds from the network in terms of bandwidth, delay, or delay jitter. For instance a voice-over-IP (VoIP) application requires both a minimum bandwidth (generally, between 20-80 Kbps, depending on the voice codec used) and a round-trip delay of about 150 ms (dictated by human ergonomics) to ensure a "good" user experience. These applications require a guarantee of service quality from the network, and the latter must reserve resources on their behalf. Furthermore, the performance that guaranteed-service applications receive is directly affected by the scheduling discipline employed by the nodes along their path, as these disciplines are responsible for scheduling packets on the outgoing links.

Based on this discussion, a packet scheduler is desirable to possess three important properties [66]:

- *Isolation and fairness.* When serving best-effort flows, it is important that the scheduler provide isolation among the competing flows and ensure that each flow receives its fair share of the link bandwidth. Isolation prevents misbehaving flows (e.g., flows transmitting too fast) from affecting other flows sharing the same link. In this context, fairness typically refers to max-min fair allocation [14] of link bandwidth among the flows, whereby flows with "small" bandwidth demands receive what they want while flows with "large" demands receive an equal share of the remaining link bandwidth.
- *Performance bounds.* Typically, guaranteed-service applications require a bandwidth bound, i.e., they must receive a minimum amount of bandwidth (measured over an appropriate interval of time). In addition, certain real-time and/or interactive applications may require bounds on packet delay. Such bounds may be expressed *deterministically* (e.g., in the form of a worst-case delay that no packet must exceed) or *statistically* (i.e., in the form of a delay threshold and a probability that any packet's delay will not exceed the threshold). Other performance bounds that have been considered include bounds on the delay jitter (defined as the difference between the largest and smallest delays experience by any packet of a flow) and on packet loss.
- *Low algorithmic complexity.* A link scheduler may need to select the next packet to serve every time a packet departs. As optical link speeds increase from a few Gbps currently to tens of Gbps and beyond, a scheduler may have only a few microseconds or less to make a decision. Hence, in order to operate at wire speeds, the scheduling discipline must be amenable to hardware implementation and require few, preferably simple, operations. In particular, since the links of backbone networks may serve hundreds of thousands of simultaneous flows, the number of operations involved in making a scheduling decision should be independent of the number of flows sharing the link.

These requirements are often contradictory. For instance, the first-come, first-served (FCFS) discipline maintains a single queue of packets and transmits them in the order of their arrival to the queue. This discipline is easy to implement in hardware and is widely deployed in routers. However, since an FCFS scheduler cannot distinguish between different flows, it cannot provide isolation among flows or guarantee per-flow performance bounds. To guarantee such bounds, a scheduler must maintain additional *scheduling state* in the form of separate queues and information regarding the requirements of each flow, which increases the complexity of its implementation.

In the following section, we review the most common packet scheduling disciplines and we discuss the tradeoffs involved with respect to the three requirements above.

10.2 Packet Scheduling Disciplines

In general, packet schedulers can be classified according to their internal structure as follows:

- *Timestamp-based schedulers* maintain a global variable, usually referred to as virtual time, to sort arriving packets and serve them in that order.
- *Frame-based schedulers* divide time into slots of fixed or variable length, and assign slots to flows in some sort of round-robin fashion.
- *Hybrid schedulers* combine features from both timestamp-based and frame-based schedulers.

Fig. 10.1 illustrates the general scheduler model that we will use in our discussion. Specifically, we assume that the scheduler serves n flows and employs per-flow queueing such that an arriving packet belonging to flow $i, i = 1, \ldots, n$, is inserted at the tail of the queue dedicated to this flow. As a result, each flow queue is sorted in increasing order of packet arrival times and its packets are served in a FCFS order. As shown in the figure, each flow i is associated with a positive real weight ϕ_i that is determined in advance (e.g., based on the application's bandwidth or delay requirements). The scheduler uses the weights in some discipline-specific manner to determine which of the head-of-line packets in the flow queues to serve next.

10.2.1 Timestamp-Based Schedulers

Timestamp schedulers emulate the ideal but unimplementable generalized processor sharing (GPS) algorithm by maintaining a virtual time function. Packets are assigned a timestamp based partly on the virtual time value at the time of their arrival, and are transmitted in increasing order of timestamp. In general, timestamp schedulers have good delay and fairness properties, but high implementation complexity, hence there has been limited deployment of such schedulers in high-speed routers.

The complexity of timestamp schedulers arises from two factors.

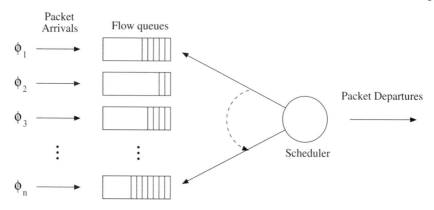

Fig. 10.1 Model of link scheduler serving n flows; ϕ_i is the weight assigned to flow i

1. *Packet sorting.* The scheduler selects among the head-of-line packets of the back-logged flows the one with the smallest timestamp to serve next; if there are n backlogged flows, this operation takes time $O(\log n)$ using a priority queue. Whereas current router technology makes it possible to support millions of flows, each with its own queue, the logarithmic complexity and the fact that the priority queue structure is not suited to hardware implementation pose significant challenges.
2. *Virtual time computation.* In order to assign a timestamp to an arriving packet, the scheduler must compute the virtual time function at the time of arrival; as we explain shortly, this computation can be expensive. One of the differentiating characteristics of timestamp scheduler variants is their use of simplified virtual time functions that are more efficient to compute.

It has also been shown that achieving a delay bound relative to GPS that is independent of the number of flows is impossible if the scheduler has a complexity below $O(\log n)$ [117].

10.2.1.1 Generalized Processor Sharing (GPS)

Generalized processor scheduling (GPS) [90] is an ideal scheduler, a theoretical construct that serves both as a starting point for designing practical scheduling disciplines and as a reference point for evaluating the fairness and delay properties of these disciplines. GPS visits each backlogged flow queue in turn and serves an infinitesimal fraction of the head-of-line packet at each queue. If flows are assigned different weights ϕ_i (refer to Fig. 10.1), then the service they receive from GPS is proportional to their weight.

If a queue is empty, GPS skips it to serve the next non-empty queue. Therefore, whenever some queues are empty, backlogged flows will receive additional service in proportion to their weights. Consequently, GPS achieves an *exact* max-min weighted fair bandwidth allocation [66]. It also provides isolation (protection) among flows, since a misbehaving flow is restricted to its fair share and does not affect other flows.

GPS is defined in a theoretical fluid flow model in which multiple queues may be served simultaneously. In a practical packet system, on the other hand, packet transmissions may not be preempted and only one queue may be served at any given time. The schedulers we describe in the following subsections attempt to emulate GPS but are designed for packetized systems.

10.2.1.2 Weighted Fair Queueing (WFQ)

Weighted fair queueing (WFQ) [28, 90] is an approximation of GPS that serves packets in the order they would complete service had they been served by GPS. Therefore, the WFQ scheduler needs to emulate the operation of the GPS server. To this end, a virtual time function $V(t)$ was proposed in [90] to track the progress of GPS. The rate of change of $V(t)$ is:

$$\frac{\vartheta V(t+\tau)}{\vartheta \tau} = \frac{1}{\sum_{i \in B(t)} \phi_i} \tag{10.1}$$

where $B(t)$ denotes the set of backlogged flows at time t and ϕ_i is the weight assigned to flow i. Let r be the rate of the link (server). In GPS, if flow i is backlogged at time t, it receives a rate of

$$\frac{\vartheta V(t+\tau)}{\vartheta \tau} \phi_i \, r = \frac{\phi_i}{\sum_{i \in B(t)} \phi_i} \, r. \tag{10.2}$$

In other words, $V(t)$ is the marginal rate at which backlogged flows receive service in GPS.

Suppose that the k-th packet of flow i arrives at time a_i^k, and has length L_i^k. Let S_i^k and F_i^k denote the virtual times at which this packet begins and completes service, respectively, under GPS. Letting $F_i^0 = 0$ for all flows i, we have [90]:

$$S_i^k = \max\{F_i^{k-1}, V(a_i^k)\} \tag{10.3}$$

$$F_i^k = S_i^k + \frac{L_i^k}{\phi_i}. \tag{10.4}$$

The WFQ scheduler serves packets in increasing order of their virtual finish times F_i^k, a policy referred to as "smallest virtual finish time first (SFF)" [11].

Let us consider the complexity of WFQ. At packet departure instants, the SFF policy is used to select the next packet to transmit. This selection can take $O(\log n)$ time, where n is the number of (backlogged) flows, if packet virtual finish times are organized in a heap-based priority queue data structure. In addition, there is the cost of maintaining the virtual time function $V(t)$ at packet arrival and departure instants. The worst-case complexity of computing $V(t)$ can be $O(n)$, although the average-case complexity is $O(1)$ [42]. Therefore, WFQ is expensive to implement within core routers that may handle hundreds of thousands to millions of flows at any given time.

The degree to which WFQ approximates GPS is determined by two properties that were established in [90]:

- *Bounded delay property.* A packet will finish service in a WFQ system no later than the time it would finish in the corresponding GPS system plus the transmission time of a maximum size packet.
- *Weak service property.* The service (in terms of total number of bits) that a flow receives in a WFQ system does not fall behind the service it would receive in the fluid GPS system by more than one maximum packet size.

While due to the second property above a WFQ system may not fall behind GPS by more than one maximum packet size, it may in fact be ahead of GPS in terms of the service provided to some flows. In particular, it was shown in [12] that WFQ may introduce substantial unfairness relative to GPS in terms of the worst-case fairness index (WFI). WFI is a metric introduced in [12] to represent the maximum time a packet arriving to an empty queue will have to wait before receiving its guaranteed service rate. Specifically, GPS has a WFI of zero, but the WFI of WFQ increases linearly with the number of flows n. Consequently, there may be substantial discrepancies in the service experienced by individual flows under the WFQ and GPS schedulers.

10.2.1.3 Worst-Case Fair Weighted Fair Queueing (WF^{2}Q)

The WF^{2}Q algorithm was introduced in [12] as a better packet approximation of GPS than WFQ. Specifically, WF^{2}Q employs a "smallest eligible virtual finish time first (SEFF)" policy for scheduling packets. A packet is eligible if its virtual start time is no greater than the current virtual time; hence, the WF^{2}Q scheduler only considers the packets that have started service in GPS to select the one to be transmitted next. It has been shown [12] that WF^{2}Q is work-conserving, maintains the bounded delay property of WFQ, and has these additional two properties:

- *Strong service property.* The service (in terms of total number of bits) that a flow receives from a WF^{2}Q system cannot fall behind (respectively, be ahead of) the service it would receive in the fluid GPS system by more than one maximum packet size (respectively, a fraction of the maximum packet size).
- *Worst-case fairness property.* The worst-case fairness index of WF^{2}Q is a constant independent of the number n of flows served by the scheduler.

The first property implies that the WF^{2}Q scheduler closely tracks the GPS system in terms of the service received by each flow, and due to the second property, WF^{2}Q is an optimal packet scheduler in terms of worst-case fairness [12].

However, the worst-case complexity of WF^{2}Q is $O(n)$, identical to that of WFQ, as both schedulers need to compute the virtual time function $V(t)$.

10.2.1.4 WF^{2}Q+

A lower-complexity scheduler, WF^{2}Q+ was introduced in [11]. The WF^{2}Q+ scheduler is work-conserving, has the same bounded delay, strong service, and worst-case fairness properties of WF^{2}Q, but uses a different virtual time function that can be computed more efficiently than the function $V(t)$ in (10.1) used by WFQ and WF^{2}Q. The new function is [11]:

$$V_{WF^2Q+}(t+\tau) = \max\left\{V_{WF^2Q+}(t)+\tau, \min_{i \in B(t)}\left\{S_i^{h_i(t)}\right\}\right\}. \qquad (10.5)$$

In the above expression, $B(t)$ is the set of backlogged flows at time t, $h_i(t)$ is the sequence number of the packet at the head of flow i's queue at time t, and $S_i^{h_i(t)}$ is the virtual start time of that packet. The minimum operation in the right-hand side of (10.5) can be performed in time $O(\log n)$ in the worst-case using a priority queue structure, hence the overall complexity of WF^{2}Q+ is $O(\log n)$, significantly lower than the $O(n)$ complexity of WFQ and WF^{2}Q.

As pointed out in [11], the WF^{2}Q+ scheduler implementation can be further simplified by maintaining a single pair of start and finish virtual time values per flow, rather than on a per-packet basis. Specifically, only a single pair of values, S_i and F_i, needs to be maintained for each flow i, corresponding to the virtual start and finish times, respectively, of the packet at the head of the queue of flow i. Let $Q_i(t-)$ denote the queue size of flow i just before time t. When a new packet reaches the head of the queue at time t, the values of S_i and F_i are updated according to the following expressions [11]:

$$S_i = \begin{cases} F_i, & Q_i(t-) \neq 0 \\ \max\{F_i, V_{WF^2Q+}(t)\}, & Q_i(t-) = 0 \end{cases} \qquad (10.6)$$

$$F_i = S_i + \frac{L_i^k}{\phi_i} \qquad (10.7)$$

where L_i^k is the length of this packet and ϕ_i is the weight assigned to the flow.

Overall, the WF^{2}Q+ scheduler achieves tight delay bounds and good worst-case fairness with a relatively low $O(\log n)$ algorithmic complexity.

10.2.1.5 Self-Clocked Fair Queueing (SCFQ)

The $O(n)$ worst-case algorithmic complexity of the WFQ and WF^{2}Q schedulers is due to the fact that the order of packet transmissions in these queueing schemes is determined by tracking the progress of the fluid-flow GPS reference system, which, in turn, requires the computation of the virtual time function $V(t)$ whose rate of change is defined in (10.1). Self-clocked fair queueing (SCFQ) [42] avoids the computationally expensive emulation of a hypothetical reference system by adopting a self-contained approach to fair queueing. Specifically, instead of using a virtual time to compute the finish times of packets as in expressions (10.3) and (10.4), SCFQ computes the finish time F_i^k of the k-th packet of flow i as:

$$F_i^k = \max\{F_i^{k-1}, F_{cur}\} + \frac{L_i^k}{\phi_i} \qquad (10.8)$$

where F_{cur} is the finish time of the packet currently in service, and finish times are initialized to $F_i^0 = 0$ for all flows i. Since the finish times can be computed in $O(1)$ time using expression (10.8), the algorithmic complexity of SCFQ is $O(\log n)$ because of the requirement to select the packet with the smallest finish time for transmission.

Although the rule (10.8) that SCFQ uses to compute packet finish times is easy to implement, the tradeoff is a much larger delay bound than WFQ. In particular, the delay bound provided by SCFQ increases linearly with the number n of flows served by the scheduler, in the worst case [44]. The worst-case fair index (WFI) of SCFQ is the same as that of WFQ, i.e, proportional to the number n of flows.

10.2.1.6 Start-Time Fair Queueing (SFQ)

Start-time fair queueing (SFQ) [45] is a variant of SCFQ that maintains both a start time and a finish time for each packet. Upon arrival, the k-th packet of flow i is assigned the start time:

$$S_i^k = \max\{F_i^{k-1}, S_{cur}\} \qquad (10.9)$$

where S_{cur} is the start time of the packet in service at the time of arrival. The finish time F_i^k of the k-th packet is computed as:

$$F_i^k = S_i^k + \frac{L_i^k}{\phi_i}. \qquad (10.10)$$

Unlike the other packet fair schedulers we have considered so far, SFQ serves packets in increasing order of their *start times*, not their finish times.

It can be seen that expressions (10.9) and (10.10) may be computed in constant time, hence SFQ has the same low algorithmic complexity $O(\log n)$ as SCFQ. However, it has been shown [45] that the worst-case delay of SFQ is significantly lower than with SCFQ. The worst-case fairness properties of SFQ are similar to those of WFQ and SCFQ.

10.2.1.7 Virtual Clock (VC)

The scheduler introduced in [119] was the first to adopt the notion of a virtual clock to represent the progress of a queueing system in terms of work (service) performed. Whereas the similar notion of virtual time has been used by fair packet queueing schedulers such as WFQ to emulate GPS, the virtual clock scheduler instead emulates time division multiplexing (TDM). Specifically, the k-th packet of flow i arriving at time t is assigned the finish time:

$$F_i^k = \max\{F_i^{k-1}, t\} + \frac{L_i^k}{\phi_i} \qquad (10.11)$$

Note that the above expression uses real time t instead of virtual time, greatly simplifying the computation of finish times. The scheduler serves packets in increasing order of their finish times, hence the complexity of virtual clock is $O(\log n)$.

Despite its simplicity, the virtual clock scheduler is able to provide delay bounds to flows. However, the use of real time t in (10.11) does not accurately represent the progress of work in the system upon the arrival of the packet. As a result, the worst-case fairness index of virtual clock can be arbitrarily large [109], even in the case of only two sessions.

10.2.2 Frame-Based Schedulers

Even with simplified virtual time computations, timestamp-based schedulers incur a substantial per-packet overhead that is related to selecting the packet with the smallest finish time to be transmitted next. Frame-based schedulers eliminate the need for packet sorting and hence achieve an $O(1)$ packet processing operation. Such schedulers typically operate by dividing time into frames. Within each frame (also referred to as round), flows are mapped to time slots of fixed or variable length and are served in a round-robin manner. Because of their low implementation complexity, frame-based schedulers have been widely deployed in high-speed routers. However, these schedulers have poor delay bound and fairness properties under most realistic traffic conditions.

10.2.2.1 Weighted Round-Robin (WRR)

The round-robin scheduler, which serves a single packet from each flow with backlogged traffic, is the simplest emulation of GPS and an early form of fair queueing. If all packets have the same size (e.g., as in ATM networks), and all flows are assigned identical weights, then round-robin is a reasonably good approximation of GPS. If flows have different weights, the more general weighted round-robin (WRR) scheduler may be used. WRR serves a number of packets from a flow in proportion to the flow's weight.

Despite its simplicity, the WRR scheduler has several limitations. In networks with variable packet sizes, emulating GPS correctly requires advance knowledge of each flow's mean packet size. Specifically, to allocate bandwidth fairly, the WRR scheduler must serve the flows according to a set of normalized weights obtained by dividing the weight of each flow by its mean packet size. In practice, however, the mean packet size for a flow may be difficult or even impossible to predict, e.g., when the flow carries compressed video whereby the packet sizes are strongly dependent on the nature of the video scenes. If mean packet sizes cannot be accurately predicted, the bandwidth allocation under WRR may be significantly different than under GPS.

A second drawback of WRR is that each backlogged flow is served exactly once within every frame. As a result, the WRR scheduler is fair only over time scales longer than the frame size: once a flow is served, it must wait for $n-1$ other flows before it gets service again. If the number n of flows is large, as is the case in core routers, this may lead to long periods of unfairness. A related issue has to do with burstiness: within each frame, a flow transmits all its packets at once, which will arrive at the downstream router in a burst. As a result, WRR has poor delay and burstiness properties.

10.2.2.2 Deficit Round-Robin (DRR)

Deficit round-robin (DRR) [106] is an improved version of WRR that makes it possible to allocate bandwidth fairly in a network with variable packet sizes without advance knowledge of each flow's mean packet size. To this end, the DRR scheduler maintains a quantum and a deficit counter for each flow that it serves. The quantum of each flow i is proportional to its weight ϕ_i. The deficit counter of each flow i is initialized to zero, and it is used to keep track of the currently unused portion of its allocated bandwidth.

Within each frame (round), the DRR scheduler visits each backlogged flow exactly once, and serves a number of packets such that the sum of their lengths does not exceed the sum of the flow's quantum and deficit counter. The value of the deficit counter is then updated to the unused portion (if any) of this latter amount, and carried over to the next round. As a result, each flow receives its fair share of the bandwidth without the need for the scheduler to estimate the mean packet size for each flow it serves.

The DRR scheduler performs a constant amount of work every time it visits the queue of a flow, its operation is easy to implement in hardware, and variants of this scheme are employed in commercial high-speed routers. However, within each frame, a flow transmits its entire quantum at once, and must then wait for all other flows to transmit before it is served again. Consequently, DRR has the same drawbacks as WRR, namely, unfairness at short time scales and poor delay and burstiness properties.

10.2.3 Hybrid Schedulers

As we have seen, the design of packet schedulers involves tradeoffs between algorithmic complexity, on the one hand, and performance bounds and fairness, on the other hand. Specifically, frame-based schedulers are simple to implement but provide poor delay bounds and suffer from short-term unfairness, while timestamp-based schedulers have good delay and fairness properties but high implementation complexity. More recently, hybrid designs have been proposed that incorporate some elements of timestamp-based schedulers into a frame-based scheme, the latter usually being a version of DRR. In doing so, hybrid schedulers attempt to combine the best of both worlds, i.e., improve the delay, fairness, and output burstiness properties of the frame-based scheduler while maintaining low implementation complexity.

10.2.3.1 Smoothed Round-Robin (SRR)

We note that the limitations of DRR in terms of short-term fairness and high output burstiness are due to the fact that the scheduler visits each flow exactly once within each frame (round), and transmits an amount of data from its queue equal to its quantum (plus the deficit counter, if a smaller amount was served in the previous round). Hence, the service each flow receives consists of long periods of inactivity (whose length is proportional to the number n of flows served by the scheduler) followed by short periods of service that result in back-to-back data transmissions (bursts). Clearly, if the service that a flow receives could be spread over the entire frame, then the fairness and output burstiness of the scheduler would be improved.

The smoothed round-robin (SRR) scheduler [25] addresses this shortcoming of DRR by spreading the quantum of service allocated to a flow over the frame using a technique based on a "weight spread sequence." The SRR scheduler needs $O(1)$ time to select a packet for transmission, and has better delay bounds than DRR. On the other hand, the worst-case delay that a packet may experience is proportional to the number n of flows served by the scheduler.

10.2.3.2 Bin Sort Fair Queueing (BSFQ)

Bin sort fair queueing (BSFQ) [23] takes a different approach to reducing the complexity of timestamp-based schedulers. In particular, BSFQ is designed to reduce the computational effort required for the two main operations of a timestamp-based scheduler: computing the virtual finish time (timestamp) of each arriving packet, and sorting packets in increasing order of timestamps. To this end, virtual time is divided into slots (bins) of length Δ, where Δ is a configurable parameter, and the scheduler maintains a virtual system clock that is equal to the left endpoint of the current slot. Arriving packets are assigned a virtual finish time using an expression similar to the one for SCFQ in (10.8), that can be computed in constant time. Packets with finish times that fall within the same slot, are inserted in a first-in, first-out (FIFO) queue associated with this slot. In other words, there is no sorting of packets that have finish times "close" to each other, as determined by the length Δ of a slot. Therefore, this "bin sorting" operation takes $O(1)$ time. When the virtual clock is equal to the left endpoint of slot i, the scheduler serves all the packets in the FIFO queue associated with slot i. When all the packets of the queue have been transmitted, the virtual clock is incremented by Δ and the scheduler serves the FIFO queue of the next slot $i + 1$.

The BSFQ scheduler is scalable and is easy to implement in hardware. Its fairness and delay guarantees depend strongly on the value of parameter Δ. When Δ is large, BSFQ reduces to FCFS, while when Δ is small, its operation is similar to that of SCFQ. While smaller Δ values result in better fairness and delay guarantees, the amount of state information that the scheduler needs to maintain increases and its efficiency decreases as the value of Δ decreases. Therefore, determining an appropriate value for Δ is a complex task that involves several tradeoffs.

10.2.3.3 Stratified Round-Robin (S-RR)

Stratified round-robin (S-RR) [93] operates by grouping ("stratifying") flows into *flow classes* based on their weights. An exponential grouping is used, such that the k-th flow class consists of flows i with weights such that: $\frac{1}{2^k} \le \phi_i < \frac{1}{2^{k-1}}$. S-RR has two scheduling components: an *intra-class* scheduler and an *inter-class* scheduler. The inter-class scheduler assigns a *scheduling interval* to each flow class such that the k-th class is assigned an interval of length 2^k slots. Within a class, flows are scheduled in the associated scheduling intervals using a variant of DRR that gives each flow a quantum that is proportional to its weight.

S-RR has low complexity and provides delay and fairness guarantees similar to those of DRR. However, it improves the worst-case delay a single packet may experience to a small constant, whereas under DRR and BSFQ this value is proportional to the number n of flows served by the scheduler.

10.2.3.4 Fair Round-Robin (FRR)

The fair round-robin (FRR) scheduler [118] is similar to S-RR in that flows are grouped into classes using the same exponential grouping, and has the same structure in that it employs both an intra-class and an inter-class scheduling component. The inter-class scheduler is timestamp-based, and determines the time a packet from each class is to be scheduled by taking into account the time-varying weight of each class (which changes over time as flows within a class become active or inactive). FRR assigns finish times to *flow classes*, not individual flows, by keeping track of the corresponding GPS system. This scheduler always serves the *eligible* flow class with the smallest finish time; eligibility of a flow class is defined as a generalization of the eligibility criterion introduced by the WF^2Q scheduler. Since the inter-class scheduler operates on the basis of flow classes, emulating GPS (i.e., computing the virtual time function) takes time proportional to the number m of classes, which, for a given system is a small constant (determined by the exponential grouping employed) that is independent of the number n of flows.

The intra-class scheduler has two functions. First, it needs to compute the class weight to pass to the inter-class scheduler; the latter uses these weights to determine the order in which each class is served, as we explained above. Second, it must decide the order in which packets from the various flows within the class will be transmitted whenever the inter-class scheduler serves this class. The intra-class scheduler uses a frame-based approach similar to DRR, but with a modification to account for the weight differences among the flows within the same class.

The FRR scheduler has $O(1)$ algorithmic complexity, is worst-case fair, and provides a single packet delay bound that is equal to a small constant, similar to S-RR.

Chapter 11
Tiered-Service Fair Queueing (TSFQ)

As we explained in the previous chapter, the high algorithmic complexity of time-stamp-based schedulers is due to two fundamental operations: (1) computation of the virtual time function to track the corresponding GPS system, and (2) packet sorting to select the packet with the smallest timestamp to serve next. The WF^2Q+ scheduler we described in Section 10.2.1.4 implements both operations in time $O(\log n)$, where n is the number of flows served. Importantly, WF^2Q+ closely emulates the ideal GPS fluid-flow scheduler, and thus it provides tight delay bounds to all flows and achieves a constant worst-case fair index. Although several timestamp-based or hybrid scheduler variants have been developed with lower complexity, these schedulers provide looser delay and fairness guarantees than WF^2Q+. As a result, the service received by individual flows under these simpler schedulers may be significantly different than the service they would receive under GPS.

We also note that the packet schedulers we reviewed in the previous chapter were designed under the assumption that both flow weights and packet sizes may take arbitrary values. This fundamental assumption underlies the high complexity of the virtual time computation and packet sorting operations. However, two important observations regarding Internet traffic characteristics suggest that the implementation of packet schedulers may be simplified significantly without compromising their delay and fairness properties.

- *Flow weights.* First, traffic flows are unlikely to have arbitrary weights. For instance, flows of guaranteed-service applications may be grouped into a small set of classes depending on the nature of the application (e.g., "voice", "video," "game,", etc) with the flows in each class having similar bandwidth and delay requirements. On the other hand, whereas best-effort applications have elastic requirements that adjust to the available rate, their bandwidth requirements are typically limited to the access bandwidth available to the user. Recall also that most Internet service providers offer some type of tiered service in which users may select only from a small set of bandwidth tiers. The practical implication of this fact is that the rates requested by flows (equivalently, the flow weights in the fair queueing system) are not arbitrary, but are limited to a small set of values that are typically known in advance. As we explain shortly, it is possible to speed

G.N. Rouskas, *Internet Tiered Services*, DOI: 10.1007/978-0-387-09738-1_11,
© Springer Science + Business Media, LLC 2009

up considerably the computation of the virtual time function if the scheduler is
designed so as to handle only a small set of discrete flow weights.

- *Packet sizes.* The second observation is that in the Internet, the vast majority
  (i.e., up to 90%) of packets have a fixed length that takes one of a small number of values [108, 112]. Therefore, the scheduler may employ simple queueing
  structures that simplify, or even completely eliminate the need for, packet sorting
  operations.

The main contribution of this chapter is a new implementation of WF^2Q+ that
exploits the above observations to ensure that the two main scheduling operations,
namely, computing the virtual time function and selecting the next packet to be
transmitted, are performed in time that is independent of the number n of flows,
while at the same time maintaining the excellent delay and fairness properties of
the original scheduler. We refer to this scheduler as *tiered-service fair queueing
(TSFQ)* [103]. In the remainder of this chapter, we first explain the operation of
TSFQ, we describe an implementation of the scheduler in the kernel of the Linux
operating system, and we present experimental performance results.

11.1 Tiered Service Fair Queueing (TSFQ)

We consider a link scheduler which serves n flows and employs per-flow queueing,
i.e., it allocates a FIFO buffer to each flow, as shown in Fig. 10.1. The scheduler
supports p distinct tiers of service, where $p \ll n$ is a small constant (e.g., $p \approx 10 -
15$). The l-th tier is characterized by a positive real weight $\phi_l, l = 1, \ldots, p$. Each
flow i is mapped to one of the p service levels, i.e., it is assigned one of the p
weights ϕ_l; we assume that this assignment remains fixed throughout the duration
of the flow. The mapping of flows to service tiers is performed at the time the flow
enters the network by taking into account the QoS requirements of the application
or the bandwidth tier to which the user subscribes. We make the assumption that
the link is configured with the number p of service tiers and the associated weights
ϕ_j. These parameters may be determined in advance by the network provider as
part of the network planning process, by using empirical information regarding the
user demands and employing the techniques we developed in earlier chapters of this
book.

Before proceeding, we emphasize that our assumption regarding flow weights
is different from the approach taken by the stratified round-robin (S-RR) [93] and
fair round-robin (FRR) [118] schedulers. Specifically, S-RR and FRR allow flows
to have *arbitrary* weights, but "stratify" them into a small number of classes using
exponential grouping. In contrast, we assume that *all flows* within a service tier are
assigned the *same* weight. To make the distinction clear, we use the term *flow tier* to
refer to a set of flows with the same weight, instead of the term *flow class* that was
used in [93, 118] to refer to a group of flows with similar weights as determined by
the specific exponential grouping method employed.

11.1.1 Logical Operation

The tiered service fair queueing (TSFQ) scheduler operates in a manner similar to WF^{2}Q+ in that:

- it uses the same virtual time function shown in expression (10.5);
- it maintains a single pair of values, S_i and F_i for each flow i, corresponding to the virtual start and finish times, respectively, of the packet at the head of the FIFO queue of flow i; these values are updated according to expressions (10.6) and (10.7); and
- it employs the SEFF policy to serve packets.

The TSFQ scheduler logically consists of two components, as illustrated in Fig. 11.1. The first component comprises of p identical *intra-tier schedulers*, while the second component is a single *inter-tier scheduler*. The main function of each component is as follows:

- *Intra-tier scheduler.* The l-th intra-tier scheduler uses the SEFF policy to select, *among the flows of the l-th service tier, $l = 1, \ldots, p$, only*, the flow i with the minimum virtual finish time F_i. The structure and operation of the intra-tier scheduler are described in detail in the following sections.
- *Inter-tier scheduler.* The inter-tier scheduler simply serves the packet at the head of the queue with the smallest virtual finish time among the p flows selected by the corresponding intra-tier schedulers. Since p is a small constant for the given link, the packet to be transmitted next can be determined in time that is independent of the number of flows, and in fact, this operation can be performed in constant time in hardware. Hence, the implementation of the inter-tier scheduler is straightforward and does not require any priority queue data structure to be maintained.

We note that the logical structure of the TSFQ scheduler is similar to the structure of the S-RR and FRR schedulers which both consist of *intra-class* and *inter-class* scheduling components. However, there are significant differences in the functionality and operation of these schedulers. Specifically, the inter-class scheduler of S-RR assigns scheduling intervals to each flow class, while the inter-class scheduler of FRR assigns weights to flow classes, not individual flows, and serves the class with the smallest timestamp. The TSFQ inter-tier scheduler, on the other hand, simply serves the flow with the smallest finish time among the p such flows across the p tiers, hence its operation is much simpler. The intra-class scheduler of S-RR serves flows within its assigned scheduling interval using a variant of DRR, while the intra-class scheduler of FRR also uses a (different) variant of DRR. In contrast, the intra-tier scheduler of TSFQ (described shortly) uses simple queueing structures to maintain the flows within its tier sorted in decreasing order of finish time.

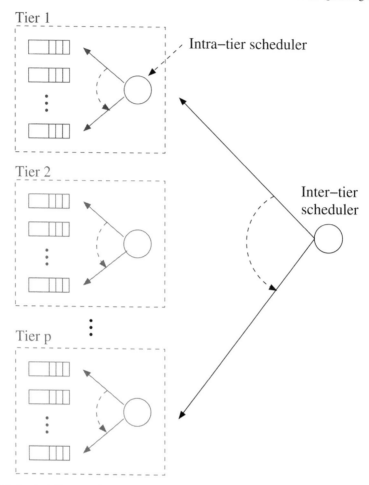

Fig. 11.1 Logical diagram of the TSFQ scheduler with p service tiers (© 2007 IEEE)

11.1.2 Virtual Time Computation

Let $S^{(l)}(t), l = 1, \cdots, p$, denote the virtual start time of the flow with the smallest finish time among the flows of the l-th service tier at time t. Then, we may rewrite the expression (10.5) of the virtual time function as:

$$V_{TSFQ}(t+\tau) = \max \left\{ V_{TSFQ}(t) + \tau, \min_{l=1,\cdots,p} \left\{ S^{(l)}(t) \right\} \right\}. \qquad (11.1)$$

Assuming that at any time t each intra-tier scheduler keeps track of the flow within its tier with the smallest finish time, the minimum operation in the right-hand side of expression (11.1) can be implemented in $O(1)$ time. Hence, the virtual time computation takes time that is independent of the number n of flows.

So far, we have shown that both the virtual time computation and the inter-tier scheduling operations take time that depends only on the number p of tiers, which is a small constant for a given scheduler. Therefore, the critical component of TSFQ is the intra-tier scheduler which is responsible for identifying (selecting) the flow with the minimum virtual finish among the flows in its tier. In the next two sections we show that by employing simple queueing structures this selection operation can be implemented efficiently.

11.2 Intra-Tier Scheduler: The Fixed-Size Packet Case

The l-th TSFQ intra-tier scheduler, $l = 1, \ldots, p$, serves flows belonging to the l-th service tier and have been assigned the same weight ϕ_l. The p intra-tier schedulers are identical and operate independently of each other. Therefore, in this and the next section we consider the operation of a single intra-tier scheduler in which all flows have identical weights. For simplicity, we let ϕ denote the weight assigned to all the flows served by the scheduler.

In this section we make the additional assumption that *all packets of all flows have constant size L* (i.e., $L_i^k = L \ \forall \ i,k$). We will remove this assumption in the next section; however, we note that the implementation we present in this section is of practical importance to ATM networks.

In a system with fixed-size packets and flows of identical weight, sorting flows according to their virtual start times produces an identical order to sorting them according to their virtual finish times. This property is formally expressed in the following lemma.

Lemma 11.1. *Consider flows i and j with $\phi_i = \phi_j = \phi$ and packets of fixed size L. Let S_i be the virtual start time of flow i, and S_j be the virtual start time of flow j. Then:*

$$S_i \leq S_j \iff F_i \leq F_j \tag{11.2}$$

Proof. TSFQ assigns finish times to flows using the expressions (10.6) and (10.7). Under the assumption of fixed packet size and identical weights, we may rewrite these expressions as:

$$S_i = \begin{cases} F_i, & Q_i(t-) \neq 0 \\ \max\{F_i, V_{TSFQ+}(t)\}, & Q_i(t-) = 0 \end{cases} \tag{11.3}$$

$$F_i = S_i + \frac{L}{\phi}. \tag{11.4}$$

Therefore, we have that:

$$S_i \leq S_j^l \iff S_i + \frac{L}{\phi} \leq S_j + \frac{L}{\phi} \iff F_i \leq F_j^l. \tag{11.5}$$

$\square$

Flow queues

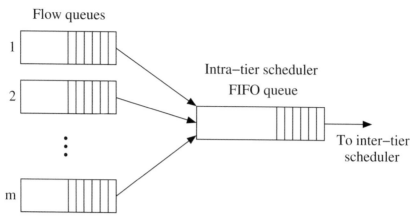

Fig. 11.2 Queue structure of the intra-tier scheduler for fixed-size packets (© 2007 IEEE)

11.2.1 Queue Structure and Operation

The intra-tier scheduler for fixed-length packets consists of a simple FIFO queue, as illustrated in Fig. 11.2. The scheduler maintains a single *token* κ_i for each flow i that it serves. Initially (i.e., at time $t = 0$, before any packet arrivals to the system), the FIFO queue is empty. Tokens are inserted at the tail of the FIFO queue representing the order in which flows will be served, and a token is removed from the head of the FIFO queue whenever it is selected for service by the inter-tier scheduler.

The operation of the scheduler is fully described by the actions taken whenever a relevant event takes place. The relevant events occur when (1) a packet arrives, (2) a flow becomes eligible for service, or (3) a packet departs (is served).

- *Packet arrival.* A packet of flow i arriving at time t is inserted at the tail of this flow's queue. If flow i was active just prior to the arrival t (i.e., its queue was non-empty, hence $Q_i(t-) \neq 0$, using the notation of Section 10.2.1.4), then no other action is taken. If, on the other hand, flow i was inactive prior to the arrival (i.e., $Q_i(t-) = 0$), then this arriving packet reaches the head of this flow's queue at time t, and the start time S_i and finish time F_i of flow i are updated according to expressions (11.3) and (11.4), respectively. If this previously inactive flow i becomes eligible at time t (i.e., $S_i \leq V_{TSFQ}(t)$ after the update), then the next event is triggered, otherwise no other action is taken.

- *A flow becomes eligible for service.* When a flow i becomes eligible at time t (i.e., $S_i = V_{TSFQ}(t)$), then the token κ_i corresponding to this flow is inserted at the tail of the scheduler's FIFO queue.

- *Packet departure.* Let κ_i be the token at the head of the scheduler's FIFO queue at the time the inter-tier scheduler selects this tier to serve. Then, the packet at the head of the queue of flow i is served and token κ_i is removed from the scheduler's FIFO queue. If flow i becomes inactive, then no other action is taken. Otherwise, a new packet reaches the head of this flow's queue, and the start time S_i and finish time F_i are updated according to expressions (11.3) and (11.4), respectively. If the flow becomes eligible, then the corresponding event above is triggered, otherwise no action is taken.

Based on these actions, it is easy to see that token κ_i is in the scheduler's FIFO queue *if and only if* flow i is eligible for service. Therefore, we have the following results.

Lemma 11.2. *Considering only the flows of a given tier, the intra-tier scheduler of Fig. 11.2 is identical to the WF^{2}Q+ scheduler [11].*

Proof. Since tokens are inserted into the FIFO queue at the moment the corresponding flows become eligible for service (i.e., at the moment their virtual start time becomes equal to the current time), tokens in the FIFO queue are sorted in increasing order of the corresponding flows' virtual start times. Because of Lemma 11.1, the queue is sorted in increasing order of the virtual finish times, which is the order in which flows are served under WF^{2}Q+. Since (1) token arrivals to the FIFO queue take place at exactly the same instants that the corresponding head-of-line packets are considered for service under WF^{2}Q+, and (2) the order of service is identical, the two schedulers are identical under the assumption of flows with fixed-size packets and identical weights. □

Lemma 11.3. *The TSFQ scheduler consisting of p intra-tier schedulers and one inter-tier scheduler is identical to WF^{2}Q+.*

Proof. Each of the p intra-tier schedulers maintains a FIFO queue that sorts the flows in its tier in increasing order of their start (equivalently, finish) times, identical to the order in which they are considered under WF^{2}Q+. The inter-tier scheduler serves the p flows with tokens at the head of the p intra-tier FIFO queues in increasing order of their virtual start (finish) times. Consequently, the TSFQ scheduler overall is identical to WF^{2}Q+. □

Based on these results and the discussion in Section 11.1.2, we conclude that the TSFQ (intra- and inter-tier) scheduler achieves the worst-case fairness and delay properties of WF^{2}Q+ with an algorithmic complexity of $O(1)$. Note that this conclusion does not contradict the findings of [117] which suggest that the $O(\log n)$ time complexity is fundamental to achieving good delay bounds. The analysis in [117] assumes that flow weights and packet sizes can take arbitrary values, whereas the result of Lemma 11.3 only holds under the specific assumptions of fixed flow weights and packet lengths.

11.3 Intra-Tier Scheduler: The Variable-Size Packet Case

We now remove the assumption we made in the previous section that all packets have a fixed size. As in the previous section, we consider the problem of scheduling flows within a given service tier, therefore we assume that all flows are assigned the same weight ϕ. In a network with variable-size packets, the statement of

Lemma 11.1 is no longer true, since the second term in the right-hand side of (11.4) is not constant. Hence, in such a network, fair queueing schedulers in general require some form of packet sorting.

In the Internet, however, it is well known that certain packet sizes dominate [108, 112]. Specifically, the study in [112] found that packets of one of three common sizes make up more than 90% of all Internet traffic; the three common packet sizes identified in the study were 40, 576, and 1500 bytes, corresponding to TCP acknowledgments, the default IP datagram size, and maximum-size Ethernet frames, respectively. A more recent study [108] shows that (1) Internet traffic is mostly bimodal at 40 and 1500 bytes, (2) there is a shift away from 576 bytes due to the proliferation of Ethernet, and (3) a new mode is forming around 1300 bytes which the authors theorize is due to widespread use of VPNs. Similar studies, which can be found on CAIDA's web site (http://www.caida.org), confirm that the length of the vast majority of Internet packets takes one of a small number of constant values. In the remainder of this section we show how we can exploit these facts regarding the Internet packet length distribution to modify the intra-tier TSFQ scheduler we presented in the previous section so that it handle Internet traffic efficiently, i.e., by performing a number of packet sorting operations that is independent of the number of flows.

11.3.1 Queue Structure and Operation

Instead of maintaining a single FIFO queue, as is the case for fixed-size packets shown in Fig. 11.2, the intra-tier scheduler for variable packet size networks maintains a small number k of queues. The queue structure of this scheduler is illustrated in Fig. 11.3 for the trimodal packet length distribution reported in [112]; the queue structure can be modified in a straightforward manner to reflect any similar distribution. In this case, the scheduler maintains $k = 7$ queues. Three of the queues are dedicated to packets of a common size, i.e., 40, 576, and 1500 bytes, respectively, which define the three modes of the distribution in [112]. The other four queues are for packets of size other the common values; as seen in Fig. 11.3, there is one queue for packets of size less than 40 bytes, one for packets of size 41-575 bytes, one for packets of size 577-1499 bytes, and one for packets of size greater than 1500 bytes.

The operation of the intra-tier scheduler is very similar to the one we described in Section 11.2.1, with only one difference. In particular, the actions taken at packet arrival and departure events are identical to those in the fixed-packet case listed in Section 11.2.1. The only difference is in the actions taken at instants when a flow becomes eligible for service:

- *A flow becomes eligible for service.* When a flow i becomes eligible at time t, then the token κ_i corresponding to this flow is inserted into the queue corresponding to the size of the packet at the head of the queue of flow i.

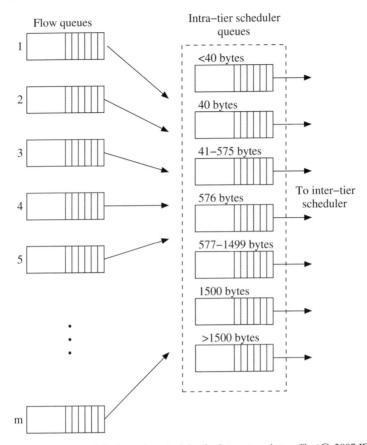

Fig. 11.3 Queue structure of the intra-tier scheduler for Internet packet traffic (© 2007 IEEE)

Similar to the fixed-packet case, token κ_i is in one of the intra-tier scheduler's queues if and only if flow i is eligible for service.

Since each of the p inter-tier schedulers maintains k distinct queues, the inter-tier scheduler selects the flow to serve next as the one with the smallest virtual finish time among the pk candidate flows whose tokens are at the head of the pk queues. Since both p and k are small integers and their values are constant for a given system, this operation of the inter-tier scheduler takes constant time, as in the fixed-size packet case.

11.3.2 Packet Sorting Operations

Note that Lemma 11.1 holds true for packets of a common size. Hence, the queues dedicated to these packets operate in a FIFO manner, and packets are simply inserted at the tail of these queues. Since packets of a common size make up more than 90% of Internet traffic [112], no sorting operations are necessary for the large majority of packets. On the other hand, queues dedicated to packets of size between the common values must be sorted appropriately at the time of a packet insertion. These sorting operations take place infrequently (less than 10% of the time), and involve relatively short queues (since less than 10% of the packets are spread over several such queues at p different service levels). Moreover, the time complexity of the sorting operations is *independent* of the number m of flows in the given service tier, and is a function only of the network load and the ratio of packets with a non-common size.

We have the following results.

Lemma 11.4. *The TSFQ scheduler for variable packet sizes, consisting of p intra-tier schedulers as in Fig. 11.3 and one inter-tier scheduler, is identical to WF^2Q+.*

Proof. The proof of Lemmas 11.2 and 11.3 also holds in this case, hence the scheduler is equivalent to WF^2Q+.   □

Finally, we note that, although consecutive packets of the same flow i may be inserted into different queues in Fig. 11.3, they will always be transmitted in order: not only does the second packet have a larger virtual finish time than the first one, but since there is exactly one token for each flow, the second packet cannot be considered for service until the first one has departed from the scheduler.

11.3.3 Elimination of Packet Sorting Operations

The operation of the intra-tier scheduler may be further simplified by eliminating packet sorting even for queues holding packets of size between the common values. Doing so may cause some packets to be served in incorrect order of virtual finish time, hence introducing a small degree of unfairness. However, the overall impact is likely to be small. Indeed, observe that packets of a non-common size represent only a small fraction of the overall traffic seen by the server, and are distributed over a number of different queues across p service tiers. Consequently, the arrival rate to each of these queues is likely to be low, especially under typical operating conditions when the load offered to the server is not too high. Now note that, since all flows within a service tier have the same weight ϕ in expression (11.4), the order of packets in such a queue will depend on the relative values of their virtual start time and length. Therefore, even when a small packet arrives to find larger packets in the queue (i.e., packets with a larger value for the second term in the right-hand side of (11.4)), the elapsed time since the previous arrival (which affects the first term of (11.4)) may be sufficiently large so that the queue remain sorted.

This intuition is further supported by the coarse manner in which the leap forward virtual clock [110] algorithm computes timestamps, and the mechanism employed by the bin sort fair queueing (BSFQ) discipline [23] to sort packets. The results in [23, 110] indicate that approximate sorting can be as good as exact sorting; moreover, in the case of our TSFQ scheduler, approximate sorting is limited to a small fraction of all packets.

We emphasize that the queue structure shown in Fig. 11.3 is for illustration purposes only and is simply meant to convey the idea underlying the structure of the scheduler for Internet packet traffic; we do not imply that routers have to be configured in exactly this manner. Network operators may configure this queue structure to reflect the specific packet distribution observed in their networks, and update it over time as traffic conditions evolve. Similarly, they may optimize the number of service tiers and the flow weights associated with them (e.g., using the techniques presented in earlier chapters) by taking into account the prevailing user demands. Therefore, this framework of fair queueing schedulers for tiered-service networks is quite flexible. Network providers may adapt the specific elements of the framework to differentiate their offerings, and to provide users with a menu of customized services.

11.4 Experimental Evaluation of TSFQ

We have developed implementations of the TSFQ scheduler for the *ns-2* network simulator and in the Linux kernel. The details of the *ns-2* implementation are reported in [16], along with a comprehensive set of simulation experiments that validate the operation of TSFQ. In this section we present network experiments with the Linux kernel implementation which is fully described in [67].

The TSFQ scheduler was implemented as a Linux kernel loadable module. The Linux kernel version 2.6.26.2 [87] was used, the latest kernel available at the time of the implementation in early 2008. The WF^2Q+ discipline [11] was also implemented as a separate loadable module for comparison purposes, since a Linux kernel implementation of the WF^2Q+ scheduler did not exist at the time..

11.4.1 Testbed and Experimental Setup

The experiments were carried out using a testbed consisting of three Linux machines connected as shown in Fig. 11.4. The leftmost machine acts as the "sender" of UDP traffic that is destined to the rightmost machine, the "receiver." The middle machine is configured as a "router" that receives packet traffic from the sender and forwards it to the receiver. The Ethernet link from the sender to the router is configured to run at 1 Gbps, while the link from the router to the receiver is configured to run at 10 Mbps. Consequently, the latter link becomes the bottleneck, causing the queues at the router to build up.

UDP traffic at the sender is generated by multiple simultaneous flows, each transmitting to a different destination port on the receiver. The router implements the

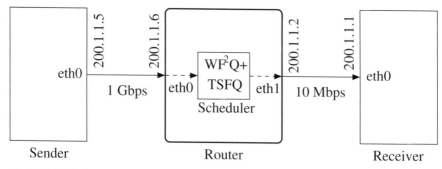

Fig. 11.4 Testbed setup

TSFQ and WF^{2}Q+ disciplines to schedule packets received from the sender for transmission on the outgoing 10 Mbps link. It also employs per-flow queueing, assigning a separate FIFO queue to each UDP flow. The router uses the port information carried by the packets to determine the flow to which they belong and insert them into the appropriate queue. The TSFQ and WF^{2}Q+ schedulers use preconfigured weights to serve the queues of the various queues.

The UDP flows at the sender continuously transmit packets to the receiver without any form of flow control. Packet sizes L are randomly generated from the following discrete distribution:

$$Pr[L=x] = \begin{cases} 0.3, x = 40 \\ 0.3, x = 1200 \\ 0.3, x = 1500 \\ 0.1, 1 \leq x \leq 39, 41 \leq x \leq 1199, 1201 \leq x \leq 1499 \end{cases} \tag{11.6}$$

This distribution generates traffic dominated by a small number (in this case, three) of packet sizes, and is similar to the packet size distributions observed in [108,112]. Consequently, the intra-tier schedulers of TSFQ are configured with six queues, similar to the structure shown in Fig. 11.3 (the seventh queue of Fig. 11.3 for packets of size greater than 1500 bytes is not used here, as no such packets are generated).

A number of experiments were carried out to investigate the behavior of the schedulers under three scenarios:

- *Scenario I.* Several flows of different weights are started at the same time. After the system reaches steady state, the flows are terminated one by one. This scenario explores how the schedulers allocate excess bandwidth to the remaining flows.
- *Scenario II.* A small number of flows are started at the same time. After the system reaches steady state, new flows of different weights are started. Once the system reaches steady state again, the newly introduced flows are terminated. The start and termination instants of the new flows are spread over time. These experiments are used to investigate the impact of new flows on the bandwidth share of existing ones, as well as the allocation of excess bandwidth.

- *Scenario III*. Many flows spanning a small number of service levels are run for a long time. This scenario is used to evaluate the fairness of each scheduling discipline. Specifically, we use Jain's fairness index [60] to compare the WF^{2}Q+ and TSFQ schedulers. In a system with n competing flows and flow i having throughput share $f_i, i = 1, \ldots, n$, Jain's fairness index (FI) is defined as:

$$FI = \frac{(\sum_{i=1}^{n} f_i)^2}{n \sum_{i=1}^{n} f_i^2}, \tag{11.7}$$

such that a value of 1 represents perfect fairness with all flows receiving an equal share $(= 1/n)$ of the available bandwidth.

11.4.2 Performance Results

In this section we present a set of illustrative experiments for the three different scenarios described above; a comprehensive suite of experimental results are available in [67].

11.4.2.1 Scenario I: Allocation of Excess Bandwidth

Figs. 11.5 and 11.6 present the results of an experiment to investigate the relative behavior of the WF^{2}Q+ and TSFQ schedulers in allocating excess bandwidth. This experiment involves four flows: flows 1 and 2 each have weight 0.15, while flows 3 and 4 each are assigned weight 0.35. For this experiment, the TSFQ scheduler was configured with $p = 2$ tiers, one with weight 0.15 and the other with weight 0.35; hence, flows 1 and 2 were assigned to the first tier, and flows 3 and 4 were assigned to the second tier. All four flows start transmission simultaneously at time $t = 0$, and are terminated one-by-one, in reverse order of their index, at 10-second intervals.

Figs. 11.5 and 11.6 plot the throughput of each flow (in Mbps) as a function of time for the WF^{2}Q+ and TSFQ schedulers, respectively. Recall that the bottleneck link in the experimental setup was set to 10 Mbps, and this latter value represents the bandwidth that is shared among the four flows. We observe that during the first 10 seconds of the experiment when all four flows are active, both schedulers allocate the available bandwidth in proportion to the flow weights, such that flows 1 and 2 (respectively, flows 3 and 4) capture approximately 15% (respectively, 35%) of the total bandwidth each. When flow 4 terminates, the bandwidth share of each of the three flows that remain active increases proportionally to its weight. In particular, flow 3 with the highest weight (.35) captures most of the bandwidth that becomes available, while flows 1 and 2 of the same but lower weight (0.15) capture an equal share of the excess bandwidth. The same behavior is observed at the time the other flows are terminated. Importantly, the throughput curves of a given flow are comparable across the two figures, implying that the TSFQ and WF^{2}Q+ schedulers perform similarly in terms of allocating bandwidth to flows in proportion to their weights.

11.4.2.2 Scenario II: Impact of New Flows

In order to demonstrate the performance of the two schedulers when flows both arrive and depart, we run an experiment with the same flows as in Scenario I, i.e., two flows of weight .35 and two of weight .15. In this case, the flows of lower weight (flows 1 and 2) both become active at time $t = 0$. At time $t = 10$ seconds (respectively, $t = 20$ seconds) flow 3 (respectively, flow 4) of higher weight becomes active. All four flows remain active until time $t = 30$ seconds, at which time flow 3 departs, followed by flow 4 at time $t = 40$ seconds.

The results of this experiment are shown in Figs. 11.7 and 11.8, which again plot the time-varying throughput of each flow under the WF^2Q+ and TSFQ scheduler, respectively. During the first ten seconds of the experiment, the two active flows receive an equal share of the available bandwidth despite their low weights, as expected. As the other two flows are introduced, the bandwidth share of existing flows is reduced accordingly; on the other hand, the bandwidth share of active flows increases as flows depart (i.e., are terminated). Overall, we make three important observations: (1) at any point in time the available bandwidth is shared among active flows in proportion to their weights; (2) as flows arrive or depart, the bandwidth share of all the flows in the system quickly reaches a new equilibrium; and (3) there is good agreement in the behavior of the two schedulers.

11.4.2.3 Scenario III: Long-Term Fairness

The first experiment of this section investigates qualitatively (i.e., graphically) the long-term fairness of the WF^2Q+ and TSFQ schedulers, and involves 32 flows that all start at time $t = 0$ and remain active throughout the duration of the experiment. Two of the flows have weight of .35, ten flows have weight equal to 0.05, and the remaining twenty flows have weight of 0.01. Hence, for this experiment, the TSFQ scheduler was configured with $p = 3$ tiers with weights of 0.35, 0.05, and 0.01, respectively, and the thirty-two flows were assigned to the appropriate tier according to their individual weights.

Figs. 11.9 and 11.10 plot the throughput of the thirty two flows as a function of time for the WF^2Q+ and TSFQ schedulers, respectively. In both figures, the flows are clearly separated in three groups, each corresponding to three TSFQ tiers, with flows within each group receiving a share of bandwidth in line with their weight. Although the throughput of the various flows shows more short-term variations under the TSFQ scheduler, the overall behavior is similar in the two figures. In order to quantify the long-term fairness of the two schedulers, we computed Jain's fairness index from expression (11.7), using the long-term throughput of the thirty two flows, and normalizing these values by the corresponding flow weight. The fairness

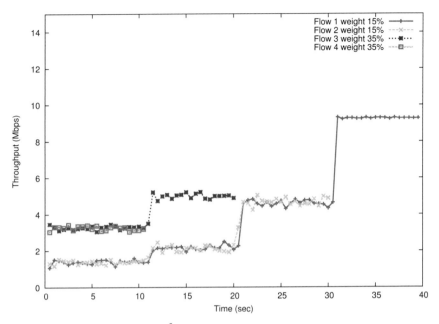

Fig. 11.5 Scenario I, four flows, WF^{2}Q+ scheduler

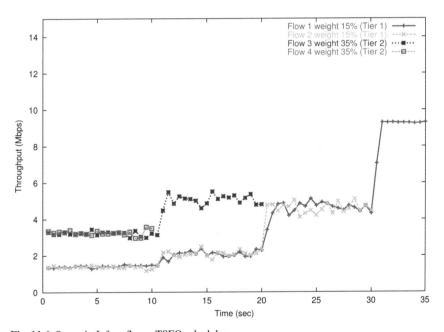

Fig. 11.6 Scenario I, four flows, TSFQ scheduler

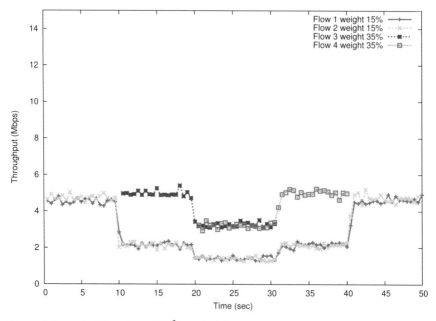

Fig. 11.7 Scenario II, four flows, WF^{2}Q+ scheduler

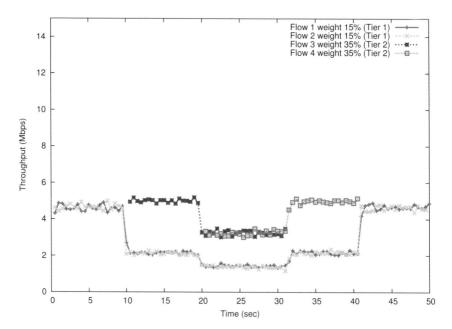

Fig. 11.8 Scenario II, four flows, TSFQ scheduler

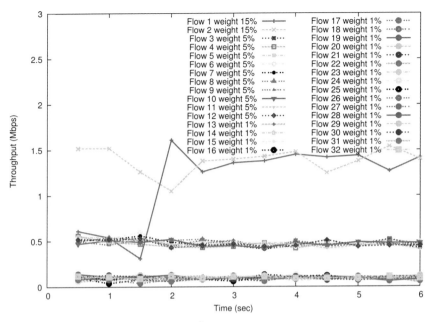

Fig. 11.9 Scenario III, thirty two flows, WF^{2}Q+ scheduler

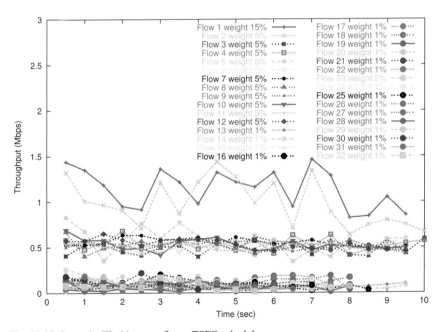

Fig. 11.10 Scenario III, thirty two flows, TSFQ scheduler

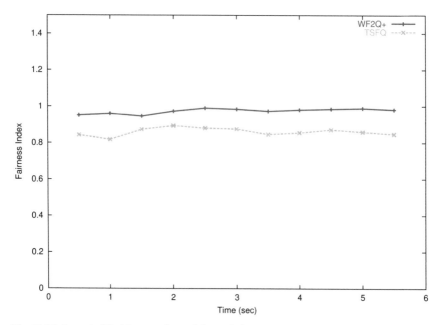

Fig. 11.11 Scenario III, thirty two flows, fairness index

index values are plotted in Fig. 11.11 as a function of time. The fairness index is about 10% higher under the WF^2Q+ scheduler, reflecting the lower throughput variations in Fig. 11.9. Nevertheless, the curves of both schedulers are relatively stable across the duration of the experiment, indicating that the two schedulers have similar fairness characteristics.

References

1. O. Aboul-Magd, L. Andersson, P. Ashwood-Smith, F. Hellstrand, K. Sundell, R. Callon, R. Dantu, L. Wu, P. Doolan, T. Worster, N. Feldman, A. Fredette, M. Girish, E. Gray, J. Halpern, J. Heinanen, T. Kilty, A. Malis, and P. Vaananen. Constraint-based LSP setup using LDP. IETF Draft <draft-ietf-mpls-cr-ldp-05.txt>, February 2001. Work in progress.
2. A. Aggarwal, M. Klawe, S. Moran, P. Shor, and R. Wilber. Geometric applications of a matrix searching algorithm. *Algorithmica*, 2(2):195–208, 1987.
3. A. Aggarwal and J. Park. Notes on searching in multidimensional monotone arrays. In *Proceedings of the 29th Annual IEEE Symposium on Foundations of Computer Science*, pages 497–512, 1988.
4. H. L. Ahuja. *Modern Microeconomics: Theory and Applications*. S. Chand and Company Ltd., 2004.
5. J. H. Anderson and A. Srinivasan. Mixed Pfair/ERfair scheduling of asynchronous periodic tasks. *Journal of Computer and System Sciences*, 68(1):157–204, February 2004.
6. D. Awduche, L. Berger, D.-H. Gan, T. Li, V. Srinivasan, and G. Swallow. RSVP-TE: Extensions to RSVP for LSP tunnels. IETF Draft <draft-ietf-mpls-rsvp-lsp-tunnel-08.txt>, February 2001. Work in progress.
7. D. O. Awduche. MPLS and traffic engineering in IP networks. *IEEE Communications*, 37(12):42–47, December 1999.
8. R. Bade and M. Parkin. *Foundations of Microeconomics*. Addison-Wesley, 2nd edition, 2004.
9. Nikhil Baradwaj. Traffic quantization and its application to QoS routing. Master's thesis, North Carolina State University, Raleigh, NC, August 2005. (2006 Graduate School Nancy G. Pollock MS Thesis Award).
10. S. K. Baruah, N. K. Cohen, C. G. Plaxton, and D. A. Varvel. Proportionate fairness: a notion of fairness in resource allocation. *Algorithmica*, 15(6):600–625, 1996.
11. J. C. R. Bennett and H. Zhang. Hierarchical packet fair queueing algorithms. In *Proceedings of ACM SIGCOMM '96*, pages 143–156, August 1996.
12. J. C. R. Bennett and H. Zhang. WF^{2}Q: worst-case fair weighted fair queueing. In *Proceedings of IEEE INFOCOM '96*, pages 120–128, 1996.
13. D. Bertsekas. *Nonlinear Programming*. Athena Scientific, 2nd edition, 2004.
14. D. Bertsekas and R. Gallager. *Data Networks*. Prentice Hall, Inc., Englewood Cliffs, NJ, 1992.
15. Dimitri P. Bertsekas. *Dynamic Programming: Deterministic and stochastic models*. Prentice-Hall, 1987.
16. Ajay Babu Amudala Bhasker. Tiered-service fair queueing (TSFQ): A practical and efficient fair queueuing algorithm. Master's thesis, North Carolina State University, Raleigh, NC, August 2006.

17. G. R. Bitran and J. C. Ferrer. On pricing and composition of bundles. *Production and Operations Management*, 16(1):93–108, February 2007.

18. P. Bonenfant and A. Rodriguez-Moral. Generic framing procedure (GFP): The catalyst for efficient data over transport. *IEEE Communications Magazine*, 40(5):72–79, May 2002.

19. Cooperative Association for Internet Data Analysis (CAIDA). Data by bytes from SDNAP traffic. http://www.caida.org/dynamic/analysis/workload/sdnap/, January 2008.

20. K. E. Case and R. C. Fair. *Principles of Economics*. Pearson Education, 7th edition, 2004.

21. D. Cavendish, K. Murakami, S-H. Yun, O. Matsuda, and M. Nishihara. New transport services for next-generation SONET/SDH systems. *IEEE Communications Magazine*, 40(5):80–87, May 2002.

22. D. Cavendish, K. Murakarni, S-H. Yun, O. Matsuda, and M. Nishihara. New transport services for next-generation SONET/SDH systems. *IEEE Communications Magazine*, 40(5):80–87, May 2002.

23. S. Cheung and C. Pencea. BSFQ: bin sort fair queueing. In *Proceedings of IEEE INFOCOM '02*, 2002.

24. A. L. Chiu and E. Modiano. Traffic grooming algorithms for reducing electronic multiplexing costs in WDM ring networks. *IEEE/OSA Journal of Lightwave Technology*, 18(1):2–12, January 2000.

25. G. Chuanxiong. SRR, an $O(1)$ time complexity packet scheduler for flows in multi-service packet networks. In *Proceedings of ACM SIGCOMM '01*, pages 211–222, August 2001.

26. M. Daskin. *Network and Discrete Location: Models, Algorithms, and Applications*. John Wiley and Sons, New York, 1995.

27. B. Davie and Y. Rekhter. *MPLS Technology and Applications*. Morgan Kaufmann Publishers, San Diego, California, 2000.

28. A. Demers, S. Keshav, and S. Shenker. Analysis and simulation of a fair queueing algorithm. In *Proceedings of ACM SIGCOMM '89*, pages 1–12, September 1989.

29. P. J. Densham and G. Rushton. A more efficient heuristic for solving large p-median problems. *Papers in Regional Science: The Journal of the RSAI*, 71(3):307–329, 1992.

30. I. Dhillon. *A New Algorithm for the Symmetric Tridiagonal Eigenvalue-Eigenvector Problem*. PhD thesis, University of California, Berkeley, 1997.

31. Z. Drezner, K. Klamroth, A. Schoebel, and G. O. Wesolowsky. The Weber problem. In *Z. Drezner and H. W. Hamacher (Editors), Facility Location: Application and Theory*, pages 1–36, Berlin, 2002. Springer Verlag.

32. E. J. Dudewicz and S. N/ Mishra. *Modern Mathematical Statistics*. John Wiley & Sons, New York, 1988.

33. R. Dutta, A. E. Kamal, and G. N. Rouskas. Grooming mechanisms in SONET/SDH and next-generation SONET/SDH. In *R. Dutta and A. E. Kamal and G. N. Rouskas (Editors), Traffic Grooming for Optical Networks: Foundations, Techniques, and Frontiers*, pages 39–55. Springer, 2008.

34. R. Dutta, A. E. Kamal, and G. N. Rouskas, editors. *Traffic Grooming in Optical Networks: Foundations, Techniques, and Frontiers*. Springer, 2008.

35. R. Dutta and G. N. Rouskas. Traffic grooming in WDM networks: Past and future. *IEEE Network*, 16(6):46–56, November/December 2002.

36. B. C. Eaton, D. F. Eaton, and D. W. Allen. *Microeconomics*. Prentice Hall, 5th edition, 2002.

37. G-S. Kuo (Ed.). Special issue on multiprotocol label switching. *IEEE Communications Magazine*, 37(12), December 1999.

38. V. Estivill-Castro and M.E. Houle. Robust distance-based clustering with applications to spatial data mining. *Algorithmica*, 30(2):216–242, 2001.

39. M. R. Garey and D. S. Johnson. *Computers and Intractability*. W. H. Freeman and Co., New York, 1979.

40. O. Gerstel, R. Ramaswami, and G. Sasaki. Cost-effective traffic grooming in WDM rings. *IEEE/ACM Transactions on Networking*, 8(5):618–630, October 2000.

41. O. Gerstel and G. Sasaki. Quality of protection: A quantitative unifying paradigm to protection service grades. In *Proceedings of SPIE Opticomm 2001*, pages 12–23, 2001.

42. S. Golestani. A self-clocked fair queueing scheme for broadband applications. In *Proceedings of IEEE INFOCOM '94*, pages 636–646, 1994.

43. W. Goralski. *SONET*. McGraw-Hill, 2000.

44. P. Goyal and H. M. Vin. Generalized guaranteed rate scheduling algorithms: A framework. *IEEE/ACM Transactions on Networking*, 5(4):561–571, August 1997.

45. P. Goyal, H. M. Vin, and H. Cheng. Start-time fair queueing: A scheduling algorithm for integrated services packet switching networks. In *Proceedings of ACM SIGCOMM '96*, pages 157–168, August 1996.

46. R. Hassin and A. Tamir. Improved complexity bounds for location problems on the real line. *Operations Research Letters*, 10:395–402, 1991.

47. L. He and J. Walrand. Pricing differentiated Internet services. In *Proceedings of INFOCOM*, 195-204 2005.

48. E. Hernandez-Valencia. Hybrid transport solutions for tdm/data networking services. *IEEE Communications Magazine*, 40(5):104–112, May 2002.

49. E. Hernandez-Valencia, M. Scholten, and Z. Zhu. The generic framing procedure (gfp): An overview. *IEEE Communications Magazine*, 40(5):63–71, May 2002.

50. C. Holahan. Time warner's net metering precedent. *Business Week*, June 4 2008.

51. C. Holahan. Time warner's pricing paradox. *Business Week*, January 18 2008.

52. C. Huang, J. Li, and K. W. Ross. Can Internet video-on-demand be profitable? In *Proceedings of ACM SIGCOMM*, pages 133–144, 2007.

53. International Telecommunication Union (ITU). Generic framing procedure (GFP). In *ITU-T G.7041*, 2001.

54. International Telecommunication Union (ITU). Link capacity adjustment scheme (LCAS) for virtually concatenated signals. In *ITU-T G.7042*, 2004.

55. L. Jackson and G. N. Rouskas. Optimal quantization of periodic task requests on multiple identical processors. *IEEE Transactions on Parallel and Distributed Systems*, 14(7):795–806, July 2003.

56. L. E. Jackson and G. N. Rouskas. Deterministic preemptive scheduling of real time tasks. *IEEE Computer*, 35(5):72–79, May 2002.

57. L. E. Jackson and G. N. Rouskas. Optimal granularity of MPLS tunnels. In *Proceedings of the Eighteenth International Teletraffic Congress (ITC 18)*, pages 1–10. Elsevier Science, September 2003.

58. L. E. Jackson, G. N. Rouskas, and M. F. M. Stallmann. The directional p-median problem: Definition, complexity, and algorithms. *European Journal of Operations Research*, 179(3):1097–1108, June 2007.

59. Laura E. Jackson. *The Directional p-Median Problem with Applications to Traffic Quantization and Multiprocessor Scheduling*. PhD thesis, North Carolina State University, Raleigh, NC, December 2003. (2004 College of Engineering Nancy G. Pollock PhD Dissertation Award).

60. R. Jain, W. Hawe, and D. M. Chiu. A quantitative measure of fairness and discrimination for resource allocation in shared systems. Technical Report TR-301, DEC Research Report, 1984.

61. H.C. Joksch. The shortest route problem with constraints. *Journal of Mathematical Analysis and Applications*, 14:191–197, 1966.

62. N. Kausar, B. Briscoe, and J. Crowcroft. A charging model for sessions on the internet. In *European Conference on Multimedia Applications, Services and Techniques*, pages 246–261, 1999.

63. F. P. Kelly. Charging and rate control for elastic traffic. *European Transactions on Telecommunications*, 8:33–37, 1997.

64. F. P. Kelly, A. Maulloo, and D. Tan. Rate control for communication networks: shadow prices, proportional fairness and stability. *Journal of Operations Research Society*, 49(3):237–252, March 1998.

65. S. Keshav. *An Engineering Approach to Computer Networking*. Addison Wesley, Reading, Massachusetts, 1997.

66. S. Keshav. *An Engineering Approach to Computer Networking.* Addison-Wesley, Reading, MA, 1997.

67. Shrikrishna Khare. Testbed implementation and performance evaluation of the tiered service fair queueing (TSFQ) packet scheduling discipline. Master's thesis, North Carolina State University, Raleigh, NC, August 2008.

68. T. D. Klastorin. The *p*-median problem for cluster analysis: A comparative test using the mixture model approach. *Management Science*, 31(1):84–95, January 1985.

69. L. Kleinrock. *Queueing Systems, Volume 1: Theory.* John Wiley & Sons, New York, 1975.

70. H. W. Kuhn. On a pair of dual nonlinear problems. In *J. Abadie (Editor), Nonlinear Programming*, pages 39–54, Amsterdam, 1967. North-Holland Publishing Company.

71. C-T. Lea and A. Alyatama. Bandwidth quantization and states reduction in the broadband ISDN. *IEEE/ACM Transactions on Networking*, 3(3):352–360, June 1995.

72. D. H. Lehmer. Mathematical models in large-scale computing units. *Annals Computing Lab HArvard University*, 26:141–146, 1951.

73. J. Y-T. Leung. A new algorithm for scheduling periodic, real time tasks. *Algorithmica*, 4:209–219, 1989.

74. J. Y-T. Leung and M. L. Merrill. A note on preemptive scheduling of periodic, real time tasks. *Information Processing Letters*, 11(3):115–118, Nov 1980.

75. D. Lichtenstein. Planar formulae and their uses. *SIAM Journal on Computing*, 11(2):329–343, April 1982.

76. L. W. McKnight and J. P. Bailey (Eds.). *Internet Economics.* MIT Press, 1997.

77. N. Megiddo and K. J. Supowit. On the complexity of some common geometric location problems. *SIAM Journal on Computing*, 13(1):182–196, February 1984.

78. D. A. Menascé, V. A. F. Almeida, and L. W. Dowdy. *Capacity Planning and Performance Modeling.* Prentice Hall, 1994.

79. E. Modiano and P. J. Lin. Traffic grooming in WDM networks. *IEEE Communications*, 39(7):124–129, Jul 2001.

80. L. Moutinho and A. Meidan. Quantitative methods in marketing. In *M. J. Baker (Editor), The Marketing Book*, pages 197–244. Butterworth-Heinemann, 2002.

81. J. F. Nash. The bargaining problem. *Econometrica*, 18:155–162, 1950.

82. J. F. Nash. Equilibrium points in *n*-person games. *Proceedings of the National Academy of Science*, 36:48–49, January 1950.

83. J. F. Nash. Two-person cooperative games. *Econometrica*, 21:128–140, 1953.

84. M. J. Neely. Optimal pricing in a free market wireless network. In *Proceedings of IEEE INFOCOM*, 2007.

85. G. P. Cornuejols G. L. Nemhauser and L. A. Wolsey. The uncapacitated facility location problem. In *P. Mirchandani and R. Francis (Editors), Discrete Location Theory*, pages 119–171. Wiley, 1990.

86. A. Odlyzko. Paris metro pricing for the Internet. In *Proceedings of the 1st ACM Conference on Electronic Commerce*, pages 140–147, 1999.

87. Linux Kernel Organization. The Linux kernel archives. http://www.kernel.org/.

88. I. H. Osman and G. Laporte. Metaheuristics: A bibliography. *Annals of Operations Research*, 63(5):511–623, October 1996.

89. L. Ostresh. The multifacility location problem: Applications and descent theorems. *Journal of Regional Science*, 17:409–419, 1977.

90. A. K. Parekh and R. G. Gallager. A generalized processor sharing approach to flow control in integrated services networks: The single-node case. *IEEE/ACM Transactions on Networking*, 1(3):344–357, June 1993.

91. S. K. Park and K. W. Miller. Random number generators: Good ones are hard to find. *Communications of the ACM*, 31(10):1192–1201, October 1988.

92. R. A. Patterson, E. Rolland, R. Barr, and G. W. Ester (Eds.). Special issue on telecommunications grooming. *Optical Networks*, 2(3), May/June 2001.

93. S. Ramabhadran and J. Pasquale. Stratified round robin: A low complexity packet scheduler with bandwidth fairness and bounded delay. In *Proceedings of ACM SIGCOMM '03*, pages 239–249, August 2003.

94. E. Rasumsen. *Games and Information: An Introduction to Game Theory.* Blackwell Publishing, 2001.

95. C. S. ReVelle. Facility siting and integer-friendly programming. *European Journal of Operational Research,* 65(2):147–158, March 1993.

96. C. S. ReVelle and R. W. Swain. Central facilities location. *Geographical Analysis,* 2:30–42, 1970.

97. D. Ros and B. Tuffin. A mathematical model of the Paris metro pricing scheme for charging packet networks. *Computer Networks,* 46(1):73–85, 2004.

98. E. Rosen, A. Viswanathan, and R. Callon. Multiprotocol label switching architecture. RFC 3031, January 2001.

99. K. E. Rosing, E. L. Hillsman, and H. Rosing-Vogelaar. A note comparing optimal and heuristic solutions to the *p*-median problem. *Geographical Analysis,* 11:86–89, 1979.

100. K. E. Rosing and C. S. ReVelle. Heuristic concentration: Two stage solution construction. *European Journal of Operational Research,* 97(1):75–86, February 1997.

101. Sheldon Ross. *Simulation (4th edition).* Elsevier Academic Press, Burlington, MA, 2006.

102. G. N. Rouskas and N. Baradwaj. A framework for tiered service in MPLS networks. In *Proceedings of IEEE INFOCOM 2007,* pages 1577–1585, May 2007.

103. G. N. Rouskas and Z. Dwekat. A practical and efficient implementation of WF^{2}Q+. In *Proceedings of IEEE ICC,* pages 172–176, June 2007.

104. D. A. Schilling, K. E. Rosing, and C. S. ReVelle. Network distance characteristics that affect computational effort in *p*-median location problems. *European Journal of Operational Research,* 127(3):525–536, December 2000.

105. S. Shenker. Fundamental design issues for the future internet. *IEEE Journal on Selected Areas in Communications,* 13(7):1176–1188, September 1995.

106. M. Shreedhar and G. Varghese. Efficient fair queueing using deficit round robin. In *Proceedings of ACM SIGCOMM '95,* 1995.

107. J. Shu and P. Varaiya. Pricing network services. In *Proceedings of IEEE INFOCOM,* pages 1221–1230, 2003.

108. R. Sinha, C. Papadopoulos, and J. Heidemann. Internet packet size distributions: Some observations. http://netweb.usc.edu/~rsinha/pkt-sizes/, October 2005.

109. D. Stiliadis and A. Varma. Latency-rate servers: A general model for analysis of traffic scheduling algorithms. *IEEE/ACM Transactions on Networking,* 6(5):611–624, October 1998.

110. S. Suri, G. Varghese, and G. Chandranmenon. Leap forward virtual clock: An $O(\log\log N)$ queueing scheme with guaranteed delays and throughput fairness. In *Proceedings of IEEE INFOCOM '97,* 1997.

111. M. B. Teitz and P. Bart. Heuristic methods for estimating the generalized vertex median of a weighted graph. *Operations Research,* 16:955–961, 1968.

112. K. Thompson, G. J. Miller, and R. Wilder. Wide-area internet traffic patterns and characteristics. *IEEE Network,* 11(6):10–23, Nov/Dec 1997.

113. P-J. Wan, G. Calinescu, L. Liu, and O. Frieder. Grooming of arbitrary traffic in SONET/WDM BLSRs. *IEEE Journal on Selected Areas in Communications,* 18(10):1995–2003, 2000.

114. H. Wang, H. Xie, L.Qiu, A. Silberschatz, and Y.R. Yang. Optimal ISP subscription for Internet multihoming: Algorithm design and implication analysis. In *Proceedings of IEEE INFOCOM,* pages 2360–2371, 2005.

115. A. Weber. *Ueber den Standort der Industrien, 1909, translated as Theory of the Location of Industries.* University of Chicago Press, Chicago, 1929.

116. A. Weinstein. *Market Segmentation.* Probus Publishing Co., 1993.

117. J. Xu and R. Lipton. On fundamental tradeoffs between delay bounds and computational complexity in packet scheduling algorithms. In *Proceedings of ACM SIGCOMM '02,* 2002.

118. X. Yuan and Z. Duan. FRR: a proportional and worst-case fair round-robin scheduler. In *Proceedings of IEEE INFOCOM '05,* March 2005.

119. L. Zhang. Virtual clock: A new traffic control scheme for packet switching networks. In *Proceedings of ACM SIGCOMM '90,* 1990.

120. K. Zhu and B. Mukherjee. Traffic grooming in an optical WDM mesh network. *IEEE Journal on Selected Areas in Communications*, 20(1):122 –133, Jan 2002.
121. K. Zhu and B. Mukherjee. A review of traffic grooming in WDM optical networks: Architectures and challenges. *Optical Networks Magazine*, 4(2):55–64, March/April 2003.

Index